The Geological Society of America, Inc.
Memoir 160

Early Proterozoic Geology
of the
Great Lakes Region

Edited by

L. G. Medaris, Jr.

1983

Published by The Geological Society of America, Inc.
3300 Penrose Place, P.O. Box 9140, Boulder, Colorado 80301
Printed in U.S.A.

Library of Congress Cataloging in Publication Data
Main entry under title:

Early proterozoic geology of the Great Lakes region.

  (Memoir / Geological Society of America ; 160)
  Bibliography: p.
  1. Geology, Stratigraphic—Pre-Cambrian—Addresses,
essays, lectures. 2. Geology—Great Lakes Region—
Addresses, essays, lectures. I. Medaris, L. G.
(L. Gordon), 1936-   . II. Series: Memoir
(Geological Society of America) ; 160.
QE653.E2   1983   551.7'15'0977   83-11509
ISBN 0-8137-1160-6

ERRATA

Geological Society of America Memoir 160
Early Proterozoic Geology of the Great Lakes Region, edited by L. G. Medaris, Jr.

Page 86 - The figure shown as Figure 2 is Figure 3 and should be on page 87.
Page 87 - The figure shown as Figure 3 is Figure 2 and should be on page 86.
          The captions for both figures are correctly placed.

Contents

Contents

Geological Society of America
Memoir 160
1983

Preface

An International Proterozoic Symposium was held at the University of Wisconsin, Madison, on May 18-21, 1981, to dedicate the recently completed Lewis G. Weeks Hall for Geological Sciences and to celebrate the centennial of the Department of Geology and Geophysics, which was originally established as the Department of Mineralogy and Geology in 1878.

The symposium topic was selected in view of the pioneering investigations of Precambrian geology in the Great Lakes region by department faculty, beginning with R. D. Irving in 1870, followed by T. C. Chamberlin, C. R. Van Hise, and C. K. Leith. Precambrian studies have continued to play an important role in departmental teaching and research over the years. For the symposium it was decided to focus on the Proterozoic Eon because at the time there were few modern, comprehensive publications available on this critical segment of Earth history and because Proterozoic events played an important role in the geological development of the Great Lakes region.

The symposium was attended by approximately 150 scientists from the United States, Canada, England, Denmark, West Germany, Australia, and the People's Republic of China. Sessions were held on Proterozoic tectonics, magmatism and metamorphism, evolution of life, atmosphere and the oceans, glaciation, mineral resources, and Proterozoic geology of the Great Lakes region. Papers from the last category are included in this volume, and papers from the other sessions will be published in a separate volume by the Geological Society of America.

Geological Society of America
Memoir 160
1983

Introduction

L. G. Medaris, Jr.
Department of Geology and Geophysics
University of Wisconsin—Madison
1215 W. Dayton St.
Madison, Wisconsin 53706

During the past ten years concerted field, laboratory, and geochronologic investigations have established a sequence of Proterozoic events for the Great Lakes region (Table 1). Although uncertainties remain with respect to the details, precise timing, and tectonic significance of some of the events, the Proterozoic history summarized in Table 1 provides a basis for further advances in deciphering the Proterozoic evolution in the region.

The papers in this volume address various early Proterozoic topics and complement the recently published Geological Society of America Memoir 156, concerned with Keweenawan geology and tectonics (Wold and Hinze, 1982). The major anorogenic magmatic event which affected much of the North American continent at approximately 1400-1500 Ma has been described in detail by Anderson (in press), Anderson and Cullers (1978), Emslie (1978), and Silver and others (1977). The Archean geology of the Great Lakes region is important in defining the geologic setting for subsequent Proterozoic development and has been discussed by Goodwin and others (1972), Morey and Hanson (1980), Morey and Sims (1976), Sims (1976), Sims and Morey (1972), Van Schmus and Anderson (1977), and Van Schmus and Bickford (1981).

The most extensive and significant event in shaping the early Proterozoic fabric of the Great Lakes region was the Penokean Orogeny, first defined by Blackwelder in 1914 and named for the Penokee Range in northern Wisconsin. Blackwelder originally assigned a post-Keweenawan age to the Penokean Orogeny, due to the apparent conformity of Upper Huronian strata and Keweenawan basalts in certain areas. The Penokean Orogeny, redefined by Goldich and others (1961) as a post-Huronian and pre-Keweenawan event, is now recognized as a major early Proterozoic orogenic belt which extends along the southern margin of the Superior Province from Minnesota to the Grenville Province, a distance of 1400 km. (Card and others, 1972; Sims, 1976; Van Schmus, 1976, 1980; Van Schmus and Bickford, 1981). The main orogenic pulse occurred at about 1800-1900 Ma, with a peak of magmatism and metamorphism at 1820-1860 Ma (Van Schmus and Bickford, 1981).

Although the spatial extent and general timing of the Penokean Orogeny have been established, there is considerable uncertainty about the tectonic processes involved in the orogeny. Models invoking an intracratonic setting (Sims, 1976; Sims and Peterman, this volume), a passive margin and subsequent plate convergence (Cambray, 1978; Larue, this volume; Van Schmus, 1976), and aulacogen development (Young, this volume) have been proposed. The type of tectonism operative during the Penokean Orogeny is important not only for documenting the geological evolution of the Great Lakes region but also for providing information on the nature of early Proterozoic tectonics, during the time of transition from Archean conditions to those of the Phanerozoic.

Table 1. Chronology of Proterozoic events in the Lake Superior region (compiled from Sims, 1976; Van Schmus, 1976, 1980; Van Schmus & Bickford, 1981; Proterozoic subdivisions following the recommendation of the International Subcommittee on Precambrian Stratigraphy, Sims, 1980)

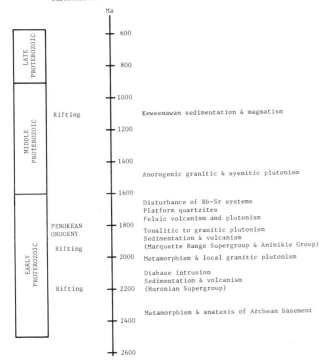

The first two papers in this volume provide an overview of the Penokean Orogeny, proposing different mechanisms for its development. The following five papers present more restricted topical or areal views of the Penokean Orogeny and are arranged geographically, from the upper peninsula of Michigan and adjacent Wisconsin, to north central Wisconsin, to Minnesota. The final two papers discuss two important post-Penokean, early Proterozoic events in the region—the 1760-Ma felsic magmatism in south central Wisconsin and the deposition and significance of red quartzites in Wisconsin and Minnesota.

It is hoped that this collection of papers will stimulate further research on remaining problems associated with the early Proterozoic development of the Great Lakes region and thereby lead to a better comprehension of the nature of early Proterozoic geologic processes.

REFERENCES CITED

Anderson, J. L., 1983, Proterozoic anorogenic granite plutonism of North America: Geological Society of America Memoir 161 (in press).

Anderson, J. L., and Cullers, R. L., 1978, Geochemistry and evolution of the Wolf River batholith, a Late Precambrian rapakivi massif in north Wisconsin, USA: Precambrian Research, v. 7, p. 287–324.

Blackwelder, E., 1914, A summary of the orogenic epochs in the geologic history of North America: Journal of Geology, v. 22, p. 633–654.

Cambray, F. W., 1978, Plate tectonics as a model for the environment of deposition and deformation of the early Proterozoic (Precambrian X) of northern Michigan [abs.]: Geological Society of America, Abstracts with Programs, v. 10, p. 376.

Card, K. D., and others, 1972, The southern province, *in* Price, R. A., and Douglas, R.J.W., eds., Variations in tectonic styles in Canada: Geological Association of Canada Special Paper 11, p. 381–433.

Emslie, R. F., 1978, Anorthosite massifs, rapakivi granites, and late Proterozoic rifting of North America: Precambrian Research, v. 7, p. 61–98.

Goldich, S. S., and others, 1961, The Precambrian geology and geochronology of Minnesota: Minnesota Geological Survey Bulletin 41, 193 p.

Goodwin, A. M., and others, 1972, The Superior province, *in* Price, R. A., and Douglas, R.J.W., eds., Variations in tectonic styles in Canada: Geological Association of Canada Special Paper 11, p. 381–433.

Larue, D. K., 1983, Three early Proterozoic terranes in the southern Lake Superior region: bearing on passive margin sedimentation and plate collision: Geological Society of America Memoir 160 (this volume).

Morey, G. B., and Hanson, G. N., 1980, Selected studies of Archean gneisses and lower Proterozoic rocks, southern Canadian shield: Geological Society of America Bulletin, v. 87, p. 141–152.

Silver, L. T., and others, 1977, The 1.4-1.5 b.y. transcontinental anorogenic plutonic perforation of North America [abs.]: Geological Society of America, Abstracts with Programs, v. 9, p. 1176–1177.

Sims, P. K., 1976, Precambrian tectonics and mineral deposits, Lake Superior region: Economic Geology, v. 71, p. 1092–1118.

——1980, Subdivision of the Proterozoic and Archean Eons: recommendations and suggestions by the International Subcommission on Precambrian Stratigraphy: Precambrian Research, v. 13, p. 379–380.

Sims, P. K., and Morey, G. B., eds., 1972, Geology of Minnesota: a centennial volume: Minnesota Geological Survey, 632 p.

Sims, P. K., and Peterman, Z. E., 1983, Evolution of Penokean foldbelt, Lake Superior region, and its tectonic environment: Geological Society of America Memoir 160 (this volume).

Van Schmus, W. R., 1976, Early and middle Proterozoic history of the Great Lakes area, North America: Philosophical Transactions of the Royal Society of London, ser. A, v. 280, p. 605–628.

Van Schmus, W. R., and Anderson, J. L., 1977, Gneiss and migmatite of Archean age in the Precambrian basement of central Wisconsin: Geology, v. 5, p. 45–48.

Van Schmus, W. R., and Bickford, M. E., 1981, Proterozoic chronology and evolution of the midcontinent region, North America, *in* Kröner, A., ed., Precambrian Plate Tectonics, Elsevier, p. 261–296.

Wold, R. J., and Hinze, W. J., 1982, Geology and tectonics of the Lake Superior basin: Geological Society of America Memoir 156, 280 p.

Young, G. M., 1983, Tectono-sedimentary history of early Proterozoic rocks of the northern Great Lakes region: Geological Society of America Memoir 160, (this volume).

Geological Society of America
Memoir 160
1983

Evolution of Penokean foldbelt, Lake Superior region, and its tectonic environment

P. K. Sims
Z. E. Peterman
U.S. Geological Survey
Denver Federal Center
Denver, CO 80225

ABSTRACT

The Penokean foldbelt is a northeast-trending zone of deformed and metamorphosed Archean and early Proterozoic rocks, as much as 250 km wide in the Lake Superior region, along the southern margin of the Superior province. The rocks within the foldbelt were deformed 1,880-1,770 m.y. ago during the Penokean tectonothermal event.

Evolution of the foldbelt began with rift-faulting that was localized along a preexisting zone of weakness in Archean rocks, the boundary between the two crustal segments previously recognized in the region: a greenstone-granite terrane (~2,700 m.y. old) to the north (Superior province) and a mostly older (in part 3,500 m.y. old) gneiss terrane to the south. The faulting and later broad foundering provided sites for deposition of detritus shed mainly from the craton to the north and for chemical sediments, including the vast iron-formations for which the region is famed. The central part of the basin also received mafic volcanic rocks intimately intercalated with the sedimentary rocks, and similar rocks dominate in the southern part. The terminal, compressional stage ended deposition. It involved folding of the supracrustal rocks, penetrative deformation of the basement rocks, and emplacement of diapiric gneiss domes with accompanying appression of intervening supracrustal rocks. Deformation was complex and prolonged, and because of basement participation it differed from place to place in style and orientation of folds and in intensity and nature of metamorphism. Granite-tonalite plutons locally were emplaced in the southern part of the foldbelt near and after the end of deformation.

The rifting and terminal compression occurred in an intracratonic or continental-margin environment; possibly the compression was caused by forces transmitted from a remote distance to the southeast, in an area undergoing rifting of a continental margin and subsequent continent-continent collision.

INTRODUCTION

This symposium is both appropriate and timely—appropriate because of the tradition of Proterozoic studies at the university, and timely because it coincides with completion of geologic reconnaissance in Wisconsin for preparation of a new million-scale geologic map of the Lake Superior region—the first since 1935.

Although reconnaissance in Wisconsin has been frustrating because of poor exposures and the absence of adequate radiometric age data, it has given us a new perspective. Mafic volcanics similar lithologically to those interbedded with graywacke turbidites (Marquette Range Supergroup) in northern Michigan increase in abundance southward and dominate the Wisconsin part of the early Proterozoic basin. In central Wisconsin, near the southern

limit of exposed Precambrian bedrock, the volcanics definitely overlie Archean basement gneisses, which have U-Pb
zircon ages of about 2,800 m.y. (Van Schmus and Anderson, 1977). An early Proterozoic age for these volcanics is
indicated by the local presence of shallow-water quartzite
and conglomerate at and near their base (Weidman, 1907).
In northern Wisconsin, there is no evidence that the volcanics have an Archean basement. In the vicinity of Wausau, in
central Wisconsin, and elsewhere, a younger succession of
apparently bimodal volcanic rocks has been distinguished.
This succession was deposited after a major episode of early
Proterozoic deformation and metamorphism related to, or
accompanied by, intense reactivation of basement gneisses.
The younger succession as well as the older rocks then were
deformed to different extents on northeast-trending fold
axes.

In this report, we describe the major aspects of the
geology of the Penokean foldbelt and present an interpretation of early Proterozoic sedimentation, volcanism, and deformation, which is a refinement of the rifting hypothesis of
Sims and others (1981).

GENERAL GEOLOGY

The broad geologic framework of the Lake Superior
region is rather well known (Sims, 1976a; Van Schmus,
1976; Sims and others, 1981). Archean basement rocks
along the southern margin of the Canadian Shield are overlain by metasedimentary and metavolcanic rocks of early
Proterozoic (Proterozoic X) age. In east-central Minnesota
and in northern Wisconsin, these rocks are intruded by
early Proterozoic granitic rocks and in turn are transected
by and partly overlain by the volcanic, gabbroic, and sedimentary rocks of the mid-Proterozoic (Proterozoic Y; Keweenawan) midcontinent rift system. The early Proterozoic
bedded rocks have been assigned to the Marquette Range
Supergroup in Michigan (Cannon and Gair, 1970) and the
Animikie and Mille Lacs Groups in Minnesota (Morey,
1973; Morey, 1978b). Possibly equivalent rocks in Wisconsin, which are dominantly mafic volcanics (Sims and others, 1978), have not been formally named.

A key element in the geology of the region, as emphasized previously (Morey and Sims, 1976; Sims and Peterman, 1981), is the existence of two contrasting Archean
crustal segments (Fig. 1). The northern part of the region is
composed of late Archean (2,750-2,600 m.y. old; Peterman,
1979) greenstone-granite complexes that are typical of
much of the Superior province. The southern part is a
complex terrane of gneisses that record a long Archean
history extending from about 3,500 m.y. ago to 2,600 m.y.
ago. The boundary between the two basement terranes
shown on Figure 1 is a modification of that in previous
reports (Sims, 1980; Sims and others, 1980; Sims and Peterman, 1981). We now interpret the boundary as having

been offset by two major northwest faults and other lesser
faults. Thus, the original northeast-trending boundary has
been stepped successively eastward by major right-lateral
wrench faults. The apparent horizontal offset of the boundary between Minnesota and Wisconsin is about 120 km;
about 70 km of this movement took place during Kween-
awan rifting. The apparent horizontal offset of the boundary by the fault passing through the Marquette trough
(near loc. 6, Fig. 1) is not known but possibly exceeds that
of the segment discussed above. The northwest faults are
part of a family of right-lateral faults of this trend in the
Lake Superior region that developed in the late Archean
and in part were rejuvenated in the Proterozoic (Sims,
1976b; Sims and others, 1978; Morey, 1978a).

The greenstone-granite complexes of the Superior
province have been described extensively. This terrane stabilized tectonically by the end of the Archean and apparently has remained close to sea level since then.

The more complex gneiss terrane in the southern part
had a long Archean history and was more mobile during
the Proterozoic than the greenstone terrane. The oldest
known rocks are approximately 3,500-m.y.-old gneisses
that occur in the Minnesota River Valley (loc. 1, Fig. 1) and
in a gneiss dome at Watersmeet, Michigan (loc. 2, Fig. 1).
In the Minnesota River Valley, detailed geochronologic
studies (Goldich and others, 1970; Goldich and others,
1980; and Goldich and Wooden, 1980) have shown that
migmatites record on Archean tectonic-igneous history extending over about 1,000 m.y. At Watersmeet, tonalitic
gneisses that have a minimum age of 3,400 m.y. (Peterman
and others, 1980) are unconformably overlain by metavolcanic rocks of late Archean (2,650 m.y. old) age (unpublished data, Peterman). Ages obtained in other parts of the
gneiss terrane are late Archean, but radiometric dating is
difficult because of widespread disturbance in the isotopic
systems. The U-Pb zircon method generally gives primary
ages, whereas Rb-Sr and K-Ar methods give much
younger, secondary ages (Van Schmus, Thurman, and Peterman, 1975). Gneisses at two widely spaced localities in
central Wisconsin (locs. 3 and 4, Fig. 1) have U-Pb zircon
ages of about 2,800 m.y. (Van Schmus and Anderson, 1977;
DuBois and Van Schmus, 1978); and the Carney Lake
Gneiss (S. B. Treves, *in* Bayley and others, 1966), near Iron
Mountain, Michigan (loc. 5, Fig. 1), has a comparable
2,800 m.y. age (Van Schmus and others, 1978).

The supracrustal metasedimentary and metavolcanic
rocks of early Proterozoic age have systematic geographic
patterns. In the north, a thin succession (~2,000 m) of clastic and chemical sedimentary rocks, including the vast iron-
formations of the Mesabi range (Bayley and James, 1973),
directly overlies greenstone-granite basement; the succession generally thickens southward over Archean gneiss
basement and is estimated to be as much as 7,500 m thick in
the Iron River district of the Menominee Range (James and

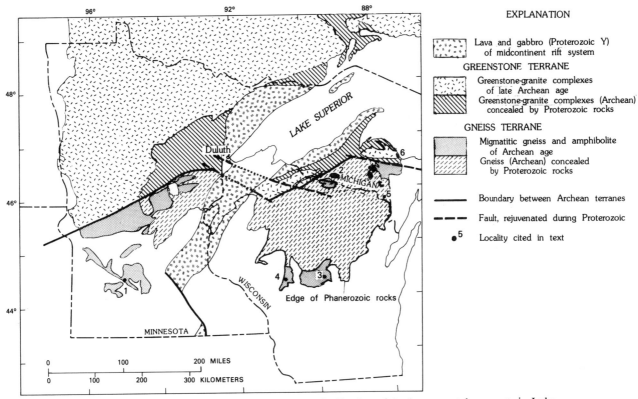

Figure 1. Map showing known and inferred distribution of Archean crustal segments in Lake Superior region. Bayfield Peninsula is underlain by Proterozoic Y sedimentary rocks. Wolf River batholith (see Fig. 2) is omitted for clarity. Revised from Sims and Peterman (1981).

others, 1968). In this, the central part of the basin, the sedimentary units in the upper part of the sequence are interbedded with submarine, mafic volcanic units that contain numerous thick sills. Further southward, in Wisconsin, volcanics compose almost all of the sequence, and quartzite, conglomerate containing granitoid clasts, and dolomite locally occur at or near their base. A younger, apparently bimodal volcanic succession with associated subvolcanic intrusive rocks (~1,860 m.y. old; Van Schmus, 1980) is present locally in Wisconsin (LaBerge, 1980; Myers and others, 1980).

Lateral and vertical variations in the early Proterozoic sequence record a complete transition from a stable craton to an unstable, deep-water environment (Bayley and James, 1973). Early deposition of quartz sands and (or) conglomerate and dolomite, probably over most of the basin, was followed by deposition of the major iron-formations of the region. This was followed by differential subsidence in the south and accumulation of a southward-thickening wedge of graywacke turbidites with intercalated, thick mafic to felsic volcanics. In the Iron River district, this clastic wedge was succeeded by an additional accumulation about 2,000 m thick of graywacke, shale, and local iron-formation.

Granitic plutons of two ages intruded the bedrock in the southern part of the region in the early Proterozoic. An older suite ranging in composition from granite to tonalite was emplaced 1,850-1,800 m.y. ago, concomitant with late stages of the Penokean tectonothermal event (Van Schmus, 1980; Morey, 1978b); and a younger suite composed dominantly of epizonal granite (Anderson and others, 1980) was emplaced about 1,760 m.y. ago (Van Schmus, 1978; 1980), after the close of the Penokean event (see Fig. 6). The largest known body of older granitic rocks, in east-central Minnesota (Morey, 1978b), is a composite batholith composed of several discrete intrusive bodies, the younger ones of which are post-tectonic. An apparently similar, but smaller pluton in northwest Wisconsin (Sims and Peterman, 1980) also is partly post-tectonic. The younger anorogenic granite occurs mainly in south-central Wisconsin in association with supracrustal rhyolite (Smith, 1978); however, granitic rocks of the same age having different compositions and fabrics have been identified by radiometric dating at scattered localities in northern Wisconsin (Van Schmus, 1980).

TECTONIC FRAMEWORK

The Penokean foldbelt is defined as the zone of deformed and metamorphosed early Proterozoic and Archean rocks adjacent to and along the southern margin of

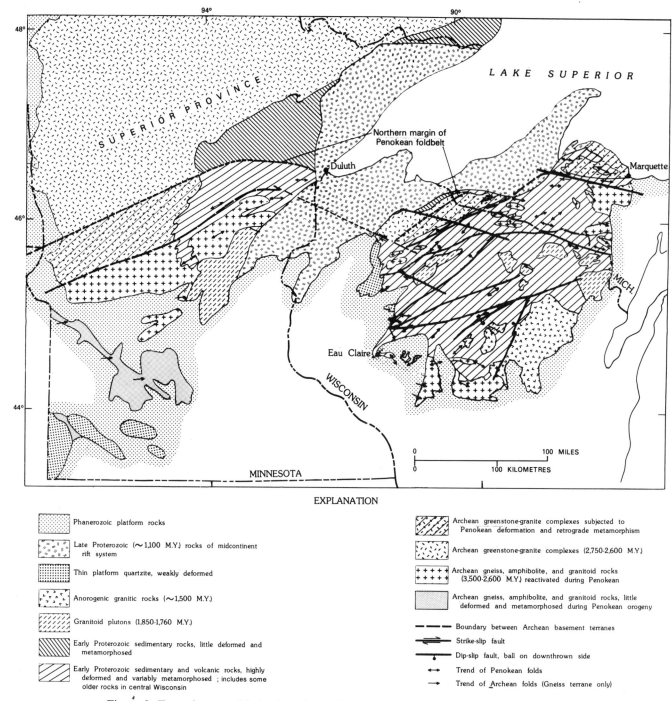

EXPLANATION

Phanerozoic platform rocks	Archean greenstone-granite complexes subjected to Penokean deformation and retrograde metamorphism
Late Proterozoic (~1,100 M.Y.) rocks of midcontinent rift system	Archean greenstone-granite complexes (2,750-2,600 M.Y.)
Thin platform quartzite, weakly deformed	Archean gneiss, amphibolite, and granitoid rocks (3,500-2,600 M.Y.) reactivated during Penokean
Anorogenic granitic rocks (~1,500 M.Y.)	Archean gneiss, amphibolite, and granitoid rocks, little deformed and metamorphosed during Penokean orogeny
Granitoid plutons (1,850-1,760 M.Y.)	Boundary between Archean basement terranes
Early Proterozoic sedimentary rocks, little deformed and metamorphosed	Strike-slip fault
Early Proterozoic sedimentary and volcanic rocks, highly deformed and variably metamorphosed; includes some older rocks in central Wisconsin	Dip-slip fault, ball on downthrown side
	Trend of Penokean folds
	Trend of Archean folds (Gneiss terrane only)

Figure 2. Tectonic map of Lake Superior region (compiled by P. K. Sims, 1981; younger succession of early Proterozoic volcanics in central Wisconsin not distinguished separately).

the Superior province (Fig. 2). The rocks within the foldbelt were deformed 1,880-1,770 m.y. ago during the Penokean tectonothermal event. The foldbelt is mainly confined to the Archean gneiss terrane, as delineated on Figure 1, but it includes an irregular 30- to 40-km-wide zone of deformed Archean and Proterozoic rocks along the southern edge of the Superior province (greenstone terrane). The

northern margin of the foldbelt, which is shown on Figure 2, is the northern limit of a Penokean penetrative foliation (tectonic front). The southern margin has not been defined. The southernmost exposures of Archean and early Proterozoic rocks in Wisconsin were deformed during the Penokean, whereas the Archean gneisses in southwestern Minnesota were not, although isotopic systems in these rocks were disturbed (Goldich and others, 1970). This definition of the foldbelt excludes the weakly deformed early Proterozoic rocks to the north of the tectonic front, which were included in the *Hudsonian foldbelt* (of the same age) on the *Tectonic map of North America* (King, 1969).

Deformation within the foldbelt was complex and prolonged and, as might be expected because of extensive participation of the basement rocks, it differed from place to place in style and orientation of folds and in intensity of metamorphism. Available field and radiometric age data are not adequate to correlate specific events over broad areas, but sequential events have been determined by careful structural studies in some areas, as in the western Marquette Range (Klasner, 1978) and the Iron River district (James and others, 1968). Additional data are presented here from central Wisconsin. In this report, we attempt to integrate broad, general features without making a detailed analysis of tectonism in space and time.

The Superior province (greenstone terrane) constituted a foreland in the stable, inner part of the Precambrian craton during the Penokean orogeny, and it acted as a rigid crustal block relative to the more mobile basement gneiss to the south. Its southern margin was deformed to different degrees and with differing styles. Along the northeast-oriented segments of the basement boundary (see Fig. 1), the rocks were intensely deformed; northwest-facing, overturned folds were developed in the early Proterozoic supracrustal rocks and a steep southeast-dipping penetrative foliation with pervasive mica and chlorite orientation was imposed on the subjacent Archean greenstone-granite complexes (see geologic sections in Sims and others, 1980, Figs. 2 and 3). On the other hand, rocks adjacent to the faulted, west-northwest-trending segments of the boundary were much less intensively deformed. Apparently, only the well-bedded sedimentary strata in the upper part of the Proterozoic sequence were folded in this environment. For example, in the western part of the Marquette Range, the Michigamme Formation is folded on east-trending axes, whereas the underlying rocks, including the Archean of the northern complex of the Marquette Range and the Amasa uplift, lack a penetrative foliation and evidence of folding (Klasner, 1978; Foose, 1981). Both Cannon (1973) and Klasner interpreted the folded Michigamme as overlying a local decollement surface.

Metamorphism of the Archean and early Proterozoic rocks along the southern margin of the Superior province mimicked the deformation. It is intense in rocks along northeast-oriented segments of the boundary, where isograds are subparallel to it and indicate steep thermal gradients (Morey, 1978a). Staurolite-grade and locally kyanite-grade metamorphism within Proterozoic rocks south of the basement boundary passes outward within a distance of 5 km or less to biotite-grade metamorphism in the vicinity of the boundary. Accordingly, metamorphism of the deformed Archean greenstone-granite basement rocks along the margin of the Superior province was retrogressive during the Penokean event, and is indicated by such features as downgraded pseudomorphs of chlorite, sericite, and biotite after garnet and staurolite. On the other hand, metamorphism of strata along the faulted, westward-trending segments of the boundary was negligible. In the western Marquette Range, metamorphism during the east-west folding was low greenschist facies. A higher grade metamorphism having an annular pattern was later superposed on these rocks during ascent of gneiss domes.

The southern part of the Penokean foldbelt, within the gneiss terrane, consists of two tectonic zones exhibiting different structural styles and metamorphic patterns. An inner zone adjacent to the basement boundary, which is a maximum of 100 km wide in northern Michigan and Wisconsin (Fig. 2), is characterized by diversely oriented folds in the Proterozoic rocks, mantled gneiss domes and uplifted Archean fault blocks, and a nodal distribution pattern of metamorphism. This zone, which contains many of the valuable iron ore deposits of the region (Bayley and James, 1973), has been recognized as the type area for the Penokean orogeny (Cannon, 1973). Adjacent to the basement boundary, the folds in the Proterozoic rocks are oriented subparallel to the boundary, overturned toward the foreland, and have a penetrative foliation marked by oriented mica and chlorite. These folds are modified adjacent to gneiss domes by superposed, younger folds that parallel the basement-cover contacts. In areas away from the gneiss domes (north of the Iron River district), the older east-trending, northward-facing fold set is truncated southward by northeast-trending folds that are overturned to the south (Cannon and Klasner, 1980). This fold set continues southward across the Iron River district of the Menominee Range (James and others, 1968) apparently into Wisconsin. Detailed studies in the Iron River district have clearly demonstrated that these northeast-trending folds were formed simultaneously with northwest-trending folds that are subparallel to the axis of the Menominee basin. The complex structural evolution in this region reflects the variable stress-strain relations caused by differential movement of basement blocks and inhomogeneous deformation of the basement. The gneiss domes and related upraised Archean fault blocks resulted from rejuvenation of the basement during the Penokean event (Sims and Peterman, 1976), and the heat attendant with their rise produced the nodal patterns of metamorphism described by James (1955).

The outer tectonic zone in the southern part of the foldbelt—in north-central and central Wisconsin—is equally as complex as the inner tectonic zone; it is composed dominantly of volcanic rocks and lacks notable iron formations. Two sets of Penokean folds have been distinguished: an older northwest- or west-trending set and a younger northeast-trending set. It is convenient to discuss

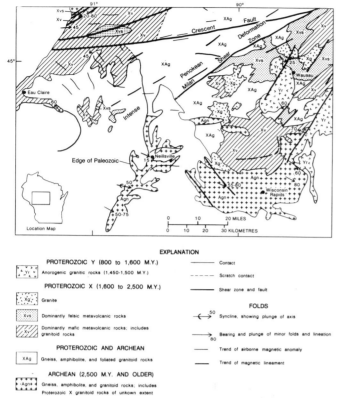

EXPLANATION

PROTEROZOIC Y (800 to 1,600 M.Y.)   ——— Contact

Anorogenic granitic rocks (1,450-1,500 M.Y.)

    - - - - - Scratch contact

PROTEROZOIC X (1,600 to 2,500 M.Y.)

    ——— Shear zone and fault

Granite

    FOLDS

Dominantly felsic metavolcanic rocks

    ←——→ 50 Syncline, showing plunge of axis

Dominantly mafic metavolcanic rocks; includes
granitoid rocks

    ——→ Bearing and plunge of minor folds and lineation
    80

PROTEROZOIC AND ARCHEAN

    ——— Trend of airborne magnetic anomaly

Gneiss, amphibolite, and foliated granitoid rocks

    ——— Trend of magnetic lineament

ARCHEAN (2,500 M.Y. AND OLDER)

Gneiss, amphibolite, and granitoid rocks; includes
Proterozoic X granitoid rocks of unkown extent

Figure 3. Generalized geologic map of central Wisconsin. Modified from LaBerge (1980) and Myers and others (1980), with additions by P. K. Sims.

the outer tectonic zone with respect to two segments or blocks that are separated by the nearly east-trending Crescent fault (Fig. 3).

The segment south of the Crescent fault (Fig. 3) contains exposed Archean as well as Proterozoic rocks. The Archean rocks were folded on northwest-trending axes in late Archean time and, where relatively undisturbed by Penokean deformation, the folds plunge moderately eastward, as at the dam at Lake Arbutus, about 30 km south of Neillsville (Maass and Van Schmus, 1980). The early Proterozoic mafic to felsic volcanics that unconformably overlie the Archean gneisses (Xv unit, Fig. 3) were folded on steep northwest axes during an early phase of the Penokean event, mainly as a result of reactivation of the basement. Concurrent with this deformation, a steep lineation was superposed on the basement gneisses and, at least locally, the older Archean structures were partly transposed (Maass and others, 1980). In an area northeast of Wisconsin Rapids (Fig. 3), mafic dikes of presumed early Proterozoic age as well as the older rocks were deformed by this deformation; they contain a steep lineation but are not folded. Following deposition of the younger, dominantly felsic volcanics in fault basins, a younger deformation produced northeast-trending folds in these supracrustal rocks and a foliation in the older rocks marked by oriented bio-

tite and chlorite. The intensity of the younger deformation varied greatly from place to place in this area. It was most intense in a N. 55-60° E.-trending, 15-km-wide zone about midway between Wausau and Eau Claire, as indicated in Figure 3; it was accompanied by amphibolite grade metamorphism. A detailed study of deformation in the Milan shear zone (Palmer, 1980), which forms the southeast margin of the zone, showed that it is characterized by a steep, northwest-dipping foliation and a rodding oriented approximately down dip. Palmer found that maximum elongation (extension) took place parallel to the steep lineation and that flattening (shortening) took place in a northwest-southeast direction, perpendicular to the foliation. Strain indicators show as much as 40 percent shortening in this zone of high strain. In other areas, as for example northeast of Eau Claire, the northeast folds warp the older foliation but do not obliterate it. The earlier phase of deformation related to widespread reactivation of the basement gneisses took place largely under T-P conditions characteristic of amphibolite grade. The younger phase, on northeast-oriented axes, was more variable and definitely lower grade in this crustal segment. The maximum metamorphism observed in the younger felsic volcanic succession is garnet grade (Myers and others, 1980). The younger volcanics at Wausau were only mildly metamorphosed; they retain delicate structures such as shards (LaBerge, 1980). Reliable radiometric dating of the younger volcanics (~1,860 m.y. old; Van Schmus, 1980) provides a maximum age for the younger phase of deformation in this area; the age of the older deformation in this area is not known.

The crustal segment north of the Crescent fault (Fig. 3) is poorly exposed except along the Chippewa River, north of Eau Claire. Definite Archean basement has not been distinguished. In this segment, the Proterozoic metavolcanic rocks (unit Xv, Fig. 3) are folded on steep east-trending axes, and magnetic anomalies indicate that this trend dominates all but the eastern part of the segment. Metamorphism of the metavolcanic rocks (Xv unit) was mainly under amphibolite-facies conditions (Myers and others, 1980).

Deciphering of Penokean structures and attendant metamorphism is complicated, especially in eastern Wisconsin and the eastern part of the Upper Peninsula, by the presence of younger, post-Penokean structures, which as yet are poorly understood. In the eastern part of the region, Rb-Sr and K-Ar dating consistently gives ages that are younger than the U-Pb zircon ages, and secondary ages in the range 1,650 to 1,600 m.y. are common (see Van Schmus, Thurman, and Peterman, 1975). Possibly this event (or events) was related to differential vertical(?) movements on preexisting faults and perhaps new fractures. It affected the 1,760-m.y.-old anorogenic granite in Wisconsin, as well as the older rocks, and is indicated in those rocks by widespread cataclasis and low-temperature

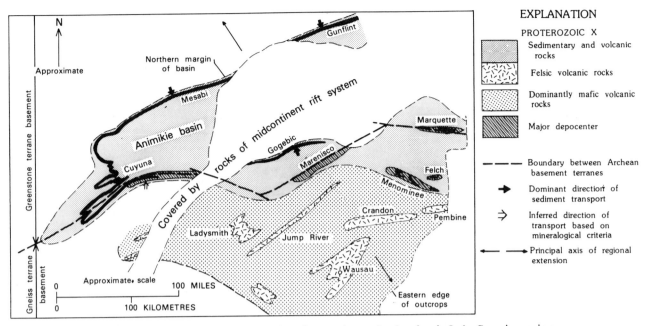

Figure 4. Patterns of early Proterozoic sedimentation and volcanism in Lake Superior region, as inferred prior to late Proterozoic (Keweenawan) rifting. Major iron-formations shown in black. Wausau shown for reference.

alteration. The 1,500-m.y.-old Wolf River batholith (Van Schmus, Medaris, and Banks, 1975) adjacent to the shear zone along its western edge (Fig. 2) has a penetrative cleavage that is subparallel to it, indicating some crustal readjustments at a still later time. Shear zones in this part of Wisconsin lie within a north-northeast-trending zone of fractures more than 800 km long that has been named (Klasner and others, 1982) the Trans Superior tectonic zone.

INTERPRETATION

An extensional stage, during which the early Proterozoic rocks were deposited, was followed shortly by a compressional stage, assigned to the Penokean tectonothermal event. The tectonic environment of the deposition and deformation has been controversial. Although a rifting environment is generally favored, there has been no consensus as to whether it was intracratonic (Sims and others, 1981) or at the margin of a continent (Van Schmus, 1976; Cambray, 1978; Larue and Glass, 1980). Also, there has been no consensus on the cause of the subsequent compression.

One of us (Sims) has favored a cratonic environment for accumulation of the early Proterozoic rocks because of the presence of Archean basement on the north, west, and south sides of the basin as well as within it; and the lack of ophiolites, blue schist assemblages, and other features characteristic of Phanerozoic plate margins. A key area for determining the tectonic setting is northern Wisconsin. In particular, the geology and geochemistry of the early Proterozoic rocks in this area need to be compared and contrasted with those in northern Michigan, known to be underlain by Archean gneisses.

Extensional Stage

Patterns of sedimentation and volcanism during the early Proterozoic and their relationships to the basement can be shown most effectively by restoring these features to their approximate positions prior to mid-Proterozoic (Keweenawan) rifting (Fig. 4). To make this reconstruction, the segment on the east side of the midcontinent rift system was shifted northwestward a distance of 70 km. This distance is the approximate width of the volcanic rocks in the midcontinent rift south of Duluth, which has been estimated as approximately the width of the opening during rifting (Chase and Gilmer, 1973).

Features not readily discernible from the geologic map are evident in the reconstruction (Fig. 4). The sedimentary rocks occupy a northeast-trending area between 200 km wide and at least 500 km long; combined sedimentation and volcanism encompassed a wider area, extending southward at least to central Wisconsin. Deposition was broadly over and along the boundary in the composite crust, with the principal depocenters being along the boundary itself and, in eastern upper Michigan, in diversely oriented basins bounded by Archean rocks.

The depositional patterns are consistent with the early Proterozoic sedimentary and volcanic rocks having been deposited in a broad depression (or cuvette) formed by rifting that was localized by the preexisting zone of weakness along the boundary between the two Archean crustal segments, but which was internally complex because of heterogeneities in the crust. Major rift-basins, including the Animikie basin, are oriented parallel to the basement boundary and approximately at right angles to the principal axis of regional extension. The Marquette basin is aligned along a major fault zone that was reactivated during the rifting and can be interpreted as a wrench-fault basin.

The clastic debris in the basin was derived mainly from the craton to the north (Morey, 1973; Morey and Ojakangas, 1970), but at least some was derived from local Archean highlands within it and from its southern flank (Peterman, 1966). An internal highland probably existed between the Gogebic and Marenisco Ranges (Fig. 4), on the presumed southeast limb of the Animikie basin, and possibly was the source of the clastic debris shed northward into the area now occupied by the Gogebic Range (Alwin, 1979). Early deposition of sediments in rift basins and adjacent, wider platform areas, as discussed by Larue and Sloss (1980), was followed by regional foundering in the south and blanketing of the sedimentary depository by a southward-thickening wedge of clastic material, in part turbidites. During and prior to deposition of this clastic wedge, the southern part of the basin received vast quantities of mafic and felsic volcanic rocks intimately intercalated with the sedimentary rocks. The patterns of sedimentation and volcanism reflect differences in the relative stabilities of the two segments of the composite Archean crust during regional stretching. The relatively stable terrane to the north—the Animikie basin—received platform type of sedimentation, whereas the more mobile terrane to the south was the site of locally thick accumulation of both sediments and volcanics.

The volcanic rocks in the southern part of the sedimentary-volcanic accumulation are thickest adjacent to the basement boundary and other presumed rift faults. They were extruded intermittently and locally are a kilometer or more thick. Those in the boundary zone, as at Marenisco (Fig. 4), are rather primitive tholeiites (unpublished data) suggesting a mantle source; whereas others have both mantle affinities and more felsic composition, suggesting partial melting of the crust.

Whether or not the younger, felsic volcanic accumulations in Wisconsin are remnants of formerly continuous volcanics of quite variable thickness, or were deposited mainly in contemporaneously fault-bounded troughs is not known; but we favor the latter. This interpretation is supported by the general parallelism of the strike of these volcanic rock units with faults bounding the basins, such as in the Wausau and Jump River areas.

In addition to the volcanic rocks, early Proterozoic mafic dikes and sills are abundant adjacent to the Archean boundary. In northern Michigan, many metamorphosed basaltic dikes occur along the northeast-trending boundary

in the vicinity of Marenisco (Sims, 1980, Fig. 5). In southwest Minnesota, mafic dikes that trend N. 50°-70° E. are common (Manzer, 1978), and a dike from one locality has a hornblende K-Ar (minimum) age of 2,080 m.y. (Hanson, 1968). This may date the approximate beginning of rifting in this part of the region. Dikes of similar orientation and composition also are present in east-central Minnesota, near St. Cloud. They have younger (~1,600-1,300 m.y.) K-Ar hornblende ages (Hanson, 1968), possibly indicating resetting by a tectonothermal event.

We have suggested (Sims and others, 1981) that rifting along the boundary (Great Lakes tectonic zone) started soon after 2,500 m.y. ago in the Lake Huron region and migrated westward with time, possibly beginning in the Lake Superior region about 2,100 m.y. ago. This pattern is consistent with the basin closing to the west in the Lake Superior region, as is suggested by the distribution of early Proterozoic bedded rocks.

It has long been recognized that the configuration of the basin in which the Animikie rocks of northern Minnesota (Morey, 1973) occur was controlled by a preexisting northeast grain in the older rocks (Van Hise and Leith, 1911; White, 1954). Rifting along the boundary of the two Archean crustal segments can account for this parallelism in trends, inasmuch as the late Archean greenstone-granite complexes that underlie the Animikie rocks are oriented subparallel to the boundary.

Compressional stage

The deformation that terminated deposition of the early Proterozoic rocks was complex because of inhomogeneities in the basement, but it mainly resulted from regional stresses that were oriented nearly perpendicular to the trends of the prevailing northeast-oriented rift-faults. The dominant direction of tectonic transport (Fig. 5), as determined from analysis of the prevailing northeast trend of folds and axial-surface foliation and from the west-northwest-trending, right-lateral wrench faults, was northwestward and probably subhorizontal. The tectonism resulted from deformation of the gneissic crust and its impingement against the more stable greenstone-granite crust of the Superior province. Deformation in the northeast-trending segments of the boundary zone took place under moderate temperature-pressure conditions, whereas that in the westward-trending segments took place under less intense conditions. As a result of local resolution of the regional stresses, basins in Michigan that had diverse orientations, such as the Menominee basin, were appressed into structural troughs. Strain was particularly high in the Archean boundary zone, along steep contacts between Archean and Proterozoic rocks, and adjacent to northeast-trending rift-faults, such as those mapped in central Wisconsin (LaBerge, 1980).

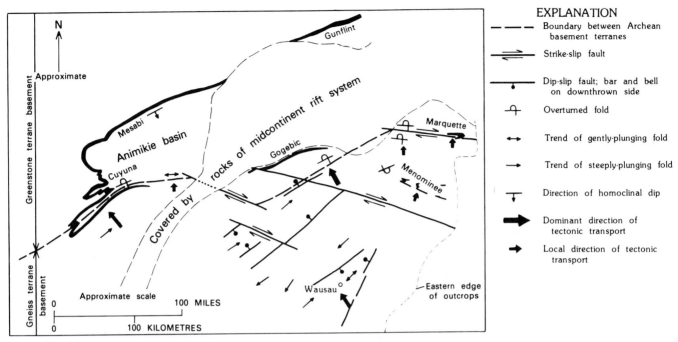

Figure 5. Regional analysis of deformation during Penokean orogeny. Same reconstruction as Figure 4.

The amount of shortening in the foldbelt as a result of the deformation is not known. Local shortening across the Milan shear zone west of Wausau was about 40 percent (Palmer, 1980), and comparable shortening probably took place across other northeast-oriented shear zones. We estimate from the shapes of folds that the shortening along the northeast-trending segments of the boundary was at least 20 percent (see Sims and others, 1980). Shortening throughout the region must have greatly exceeded the expansion resulting from dilation by the voluminous mafic dikes emplaced in the vicinity of the basement boundary.

Deformation related to reactivation of the basement, such as the gneiss domes, was accompanied by some shortening—mainly, the bedded rocks in narrow troughs were tightly appressed. In the complex area of the Menominee Range and Felch trough (see Bayley and others, 1966), it is evident that reactivation of the basement rocks in northeastern Wisconsin and their expansion northward against the rigid basement to the north produced high-angle reverse faults in the intervening basement as well as tight folds in the supracrustal Proterozoic rocks. The iron-bearing rocks of the Menominee Range were repeated by thrust faults, and the basement rocks were broken into a jumble of fault-bounded blocks (see Dutton and Linebaugh, 1967). The largest of the faults, named the Bush Lake fault, marks the northern limit of this complex area and juxtaposes east-trending structural blocks against a northwest-trending structural entity (James and others, 1961).

The Archean basement rocks were deformed during the Penokean tectonism by both ductile and brittle deformation. Most of the intense deformation was accomplished by movement on minute fractures and along grain boundaries, which resulted in grain-size reduction and little change in mineralogy. Cataclasis and recrystallization of granitoid rocks was accompanied or followed by differential mobility of potassium and probably other lithophile elements, resulting in open systems that plague our attempts to date metamorphosed rocks in the region.

CONCLUSIONS AND IMPLICATIONS

A scenario consisting of rifting followed by compression in an intracratonic or, possibly, a continental-margin setting can best account for the patterns of deposition and deformation in the Lake Superior region during early Proterozoic time. A summary of the events through time is shown in Figure 6. The apparent axis of both rifting and maximum shortening during the Penokean tectonism was northwest-southeast, approximately perpendicular to the orientation of the boundary in the composite Archean crust.

The principal problem in interpreting the tectonism is to reconcile a pattern of stratigraphy, sedimentology, structure, and magmatism that argues for an extensional tectonic regime that is followed relatively soon afterwards by strong compression. Passive mantle upwelling involving either plumes or convection currents conceivably could account for the extension. The observed tectonism in the region could have resulted from forces transmitted from a remote distance to the southeast, in an area undergoing rifting of a continental margin and subsequent continent-continent collision. This mechanism would require a continuous basement of Archean or older Proterozoic rocks extending to this continental margin, for which there is no supporting evidence at present. It may be significant that the buried trace of the Grenville Front tectonic zone in the eastern United States and the opening that created Iapetus at about the beginning of Phanerozoic time are subparallel to the prevalent Penokean structures. Was it possible that a proto-Atlantic ocean was formed in the early Proterozoic?

By reference to Figure 4, it can be seen that the axis of the middle Proterozoic (Keweenawan) midcontinent rift system in western Lake Superior coincides approximately with the zone of earlier rifting. The close spatial mimicry of these patterns in this region suggests that they are a part of a tectonic continuum that began with the sequential development of the composite Archean crust and culminated at two later stages, with tectonothermal pulses during the Penokean event and during the later Keweenawan (~1,100 m.y. ago) rifting. Lesser extensional and compressional pulses occurred during and after the major episodes of tectonism (Sims and others, 1980; Sims and others, 1981).

A model for the development and evolution of sedimentary basins recently proposed by McKenzie (1978) can account for many of the observed features in the Penokean foldbelt. In McKenzie's model, rapid stretching of the continental lithosphere and crust produces thinning and passive upwelling of hot asthenosphere. This stage, which is associated with block faulting and subsidence, is followed by thickening of the lithosphere by heat conduction to the surface and further, broad foundering. The magnitude of the subsidence and the time involved are dependent on the amount of stretching and the original thickness of the lithosphere. The crustal expression of the stretching in the Lake Superior region is the rift-faulting, which provided depositories for sediment accumulation and a mechanism to trigger volcanism. Unlike plate convergent environments, where sedimentary and volcanic accumulations can be very great through development of accretionary wedges, the accumulations in this model are limited in thickness and directly related to the amount of thinning. The maximum thickness of the infilling (8 km or less) in the region is consistent with a calculated maximum for a basin of about this age when stretched by a factor of two (McKenzie and others, 1980). The thermal perturbation accompanying the thinning can explain the greater plasticity of the Archean gneiss basement during tectonism and the generation of granitic magmas within this terrane.

Subsequent shortening of the basin during the compressional stage was accommodated to a large extent by ductile shear in contact zones and by penetrative deformation in the gneiss basement and overlying supracrustal sequences. Shortening across the foldbelt was locally sub-

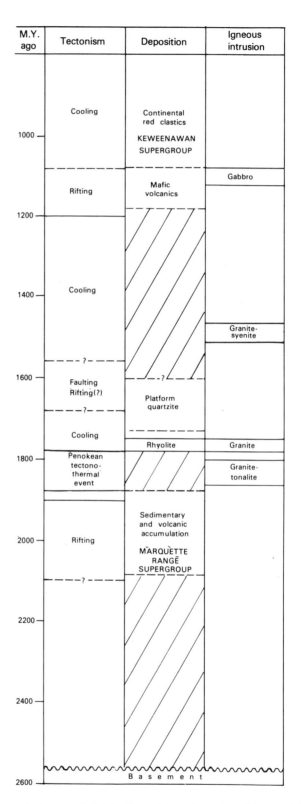

Figure 6. Temporal distribution of tectonism, deposition, and igneous intrusion in Penokean foldbelt.

stantial. Formation of the basin by lithospheric distension requires the corollary of crustal thickening during the compressional stage. Reestablishment of a "normal" crustal thickness would require shortening in the range of 20-30 percent. Greater shortening would produce an overthickened crust. Uplift and erosion would result in the removal of part of the folded supracrustals, leaving remnants in the deeper fault-bounded basins.

ACKNOWLEDGMENTS

This report is an outgrowth of continuing field and laboratory studies of the Precambrian rocks in the Lake Superior region, carried out in conjunction with individuals of the U.S. Geological Survey, the Minnesota Geological Survey, and the Wisconsin Geological and Natural History Survey since 1973. Knowledge of the geology of the Lake Superior region has been cumulative over the past several decades, but we would particularly like to acknowledge the contributions of H. L. James and colleagues of the U.S. Geological Survey, G. B. Morey of the Minnesota Geological Survey, and W. R. Van Schmus of the University of Kansas. Recent geologic mapping sponsored by the Wisconsin Geological and Natural History Survey has been helpful. M. G. Mudrey, Jr., provided invaluable assistance during our reconnaissance in northern Wisconsin, and P. E. Myers, G. L. LaBerge, W. F. Cannon, and R. S. Maass provided many stimulating discussions on the outcrops. However, we assume full responsibility for the conclusions reported here. Critical reviews by W. F. Cannon, J. S. Klasner, R. H. Moench, and Grant Young improved the manuscript.

REFERENCES CITED

Alwin, Bevan, 1979, Sedimentology of the Tyler Formation [abs.]: Geological Society of America Abstracts with Programs, v. 11, no. 5, p. 225.

Anderson, J. L., Cullers, R. L., and Van Schmus, W. R., 1980, Anorogenic metaluminous and peraluminous granite plutonism in the mid-Proterozoic of Wisconsin, U.S.A.: Contributions to Mineralogy and Petrology, v. 74, p. 311–328.

Bayley, R. W., Dutton, C. E., and Lamey, C. A., 1966, Geology of the Menominee iron-bearing district, Dickinson County, Michigan, and Florence and Marinette Counties, Wisconsin: U.S. Geological Survey Professional Paper 513, 96 p.

Bayley, R. W., and James, H. L., 1973, Precambrian iron-formations of the United States: Economic Geology, v. 68, p. 934–959.

Cambray, F. W., 1978, Plate tectonics as a model for the environment of deposition and deformation of the early Proterozoic (Precambrian X) of northern Michigan [abs.]: Geological Society of America Abstracts with Programs, v. 10, no. 7, p. 376.

Cannon, W. F., 1973, The Penokean orogeny in northern Michigan, *in* Young, G. M., ed., Huronian stratigraphy and sedimentation: Geological Association of Canada Special Paper 12, p. 211–249.

Cannon, W. F., and Gair, J. E., 1970, A revision of stratigraphic nomenclature for middle Precambrian rocks in northern Michigan: Geological Society of America Bulletin, v. 81, p. 2843–2846.

Cannon, W. R., and Klasner, J. S., 1980, Bedrock geologic map of Kenton-Perch Lake area, northern Michigan: U.S. Geological Survey Miscellaneous Investigations Map I-1290 (scale 1:24,000).

Chase, C. G., and Gilmer, T. H., 1973, Precambrian plate tectonics—The midcontinent gravity high: Earth and Planetary Science Letters, v. 21, p. 70–78.

DuBois, J. F., and Van Schmus, W. R., 1978, Petrology and geochronology of Archean gneiss in the Lake Arbutus area, west-central Wisconsin [abs.]: Abstracts and Proceedings, 24th Institute on Lake Superior Geology, Milwaukee, Wisconsin, p. 11.

Dutton, C. E., and Linebaugh, R. E., 1967, Map showing Precambrian geology of the Menominee iron-bearing district and vicinity, Michigan and Wisconsin: U.S. Geological Survey Miscellaneous Geologic Investigations Map I-466 (scale 1:125,000).

Foose, M. P., 1981, Geologic map of the Ned Lake quadrangle, Iron and Baraga Counties, Michigan: U.S. Geological Survey Miscellaneous Investigations Series Map I-1284.

Goldich, S. S., Hedge, C. E., and Stern, T. W., 1970, Age of the Morton and Montevideo gneisses and related rocks, southwestern Minnesota: Geological Society of America Bulletin, v. 81, p. 3671–3696.

Goldich, S. S., Hedge, C. E., Stern, T. W., Wooden, J. L., Bodkin, J. B., and North, R. M., 1980, Archean rocks of the Granite Falls area, southwestern Minnesota: Geological Society of America Special Paper 182, p. 19–43.

Goldich, S. S., and Wooden, J. L., 1980, Origin of the Morton Gneiss, southwestern Minnesota, part III, Geochronology: Geological Society of America Special Paper 182, p. 77–94.

Hanson, G. N., 1968, K-Ar ages for hornblende from granites and gneisses and for basaltic intrusives in Minnesota: Minnesota Geological Survey Report of Investigations 8, 20 p.

James, H. L., 1955, Zones of regional metamorphism in the Precambrian of northern Michigan: Geological Society of America Bulletin, v. 66, p. 1455–1487.

James, H. L., Clark, L. D., Lamey, C. A., and Pettijohn, F. J., 1961, Geology of central Dickinson County, Michigan: U.S. Geological Survey Professional Paper 310, 176 p.

James, H. L., Dutton, C. E., Pettijohn, F. J., and Wier, K. L., 1968, Geology and ore deposits of the Iron River-Crystal Falls district, Iron County, Michigan: U.S. Geological Survey Professional Paper 570, 134 p.

King, P. B., 1969, Tectonic map of North America: U.S. Geological Survey (scale 1:5,000,000).

Klasner, J. S., 1978, Penokean deformation and associated metamorphism in the western Marquette Range, northern Michigan: Geological Society of America Bulletin, v. 89, p. 711–722.

Klasner, J. S., Cannon, W. F., and Van Schmus, W. R., 1982, The pre-Keweenawan tectonic history of southern Canadian Shield and its influence on formation of the Midcontinent Rift: Geological Society of America Memoir 156, p. 27–46.

LaBerge, G. L., 1980, The Precambrian geology and tectonics of Marathon County, Wisconsin: Field Trip Guidebook 4, 26th Annual Institute on Lake Superior Geology, Eau Claire, Wisconsin, p. 1–32.

Larue, D. K., and Sloss, L. L., 1980, Early Proterozoic sedimentary basins of the Lake Superior region—Summary: Geological Society of America Bulletin, part I, v. 91, p. 450–452.

Maass, R. S., Medaris, L. G., Jr., and Van Schmus, W. R., 1980, Penokean deformation in central Wisconsin: Geological Society of America Special Paper 182, p. 147–157.

Maass, R. S., and Van Schmus, W. R., 1980, Precambrian tectonic history of the Black River valley: Field Trip Guidebook 2, 26th Annual Institute on Lake Superior Geology, Eau Claire, Wisconsin, 43 p.

Manzer, G. K., 1978, Petrology and geochemistry of Precambrian mafic dikes, Minnesota, and their bearing on the secular chemical variations in Precambrian basaltic magmas [Unpublished Ph.D. Thesis]: Rice University, Houston Texas, 217 p. (typescript).

McKenzie, D. P., 1978, Some remarks on the development of sedimentary basins: Earth and Planetary Science Letters, v. 40, p. 25–32.

McKenzie, D. P., Nisbet, Euan, Sclater, J. G., 1980, Sedimentary basin development in the Archean: Earth and Planetary Science Letters, v. 48, p. 35–41.

Morey, G. B., 1973, Stratigraphic framework of Middle Precambrian rocks in Minnesota, *in* Young, G. M., ed., Huronian stratigraphy and sedimentation: Geological Association of Canada Special Paper 12, p. 211–249.

——1978a, Metamorphism in the Lake Superior region, U.S.A., and its relation to crustal evolution, *in* Metamorphism in the Canadian Shield: Geological Survey of Canada Paper 78–10, p. 283–314.

——1978b, Lower and middle Precambrian stratigraphic nomenclature for east-central Minnesota: Minnesota Geological Survey Report of Investigations 21, 52 p.

Morey, G. B., and Ojakangas, R. W., 1970, Sedimentology of the middle Precambrian Thomson Formation, east-central Minnesota: Minnesota Geological Survey Report of Investigations 13, 32 p.

Morey, G. B., and Sims, P. K., 1976, Boundary between two Precambrian W terranes in Minnesota and its geologic significance: Geological Society of America Bulletin, v. 87, p. 141–152.

Myers, P. E., Cummings, M. L., and Wurdinger, S. R., 1980, Precambrian geology of the Chippewa Valley: Field Trip Guidebook 1, 26th Annual Institute on Lake Superior Geology, Eau Claire, Wisconsin, 123 p.

Palmer, E. A., 1980, The structure and petrology of Precambrian metamorphic rock units, northwestern Marathon County, Wisconsin [Unpublished M.S. thesis]: University of Minnesota-Duluth, Duluth, Minnesota, 127 p.

Peterman, Z. E., 1966, Rb-Sr dating of middle Precambrian metasedimentary rocks of Minnesota: Geological Society of America Bulletin, v. 77, no. 10, p. 1031–1043.

——1979, Geochronology and the Archean of the United States: Economic Geology, v. 74, p. 1544–1562.

Peterman, Z. E., Zartman, R. E., and Sims, P. K., 1980, Early Archean tonalitic gneiss from northern Michigan, U.S.A.: Geological Society of America Special Paper 182, p. 125–134.

Sims, P. K., 1976a, Precambrian tectonics and mineral deposits, Lake Superior region: Economic Geology, v. 71, p. 1092–1118.

——1976b, Early Precambrian tectonic-igneous evolution in the Vermilion district, northeastern Minnesota: Geological Society of America Bulletin, v. 87, p. 379–389.

——1980, Boundary between Archean greenstone and gneiss terranes in northern Wisconsin and Michigan: Geological Society of America Special Paper 182, p. 113–124.

Sims, P. K., Cannon, W. R., and Mudrey, M. G., Jr., 1978, Preliminary geologic map of Precambrian rocks in northern Wisconsin: U.S. Geological Survey Open-File Report 78-318 (scale 1:250,000).

Sims, P. K., Card, K. D., and Lumbers, S. B., 1981, Evolution of early Proterozoic basins of the Great Lakes region. *in* Proterozoic basins of Canada, F.H.A. Campbell, ed.: Geological Survey of Canada Paper 81-10, p. 379–397.

Sims, P. K., Card, K. D., Morey, G. B., and Peterman, Z. E., 1980, The Great Lakes tectonic zone—a major crustal structure in central North America: Geological Society of America Bulletin, v. 91, p. 690–698.

Sims, P. K., and Peterman, Z. E., 1976, Geology and Rb-Sr ages of reactivated Precambrian gneiss and granite in the Marenisco-Watersmeet area, northern Michigan: U.S. Geological Survey Journal of Research, v. 4, no. 4, p. 405–414.

——1980, Geology and Rb-Sr age of lower Proterozoic granitic rocks, northern Wisconsin: Geological Society of America Special Paper 182, p. 139–146.

——1981, Archean rocks in the southern part of the Canadian Shield—A review: Second International Archean Symposium, Perth, 1980, J. E. Glover and D. I. Groves, eds., Geological Society of Australia Special Publication 7, p. 85–98.

Smith, E. I., 1978, Precambrian rhyolites and granites in south-central Wisconsin—Field relations and geochemistry: Geological Society of America Bulletin, v. 89, p. 875–890.

Van Hise, C. R., and Leith, C. K., 1911, The geology of the Lake Superior region: U.S. Geological Survey Monograph 52, 641 p.

Van Schmus, W. R., 1976, Early and middle Proterozoic history of the Great Lakes area, North America: Royal Society of London Philosophical Transactions, ser. A280, no. 1298, p. 605–628.

——1978, Geochronology of the southern Wisconsin rhyolites and granites: Geoscience Wisconsin, v. 2, p. 19–24.

——1980, Chronology of igneous rocks associated with the Penokean orogeny in Wisconsin: Geological Society of America Special Paper 182, p. 159–168.

Van Schmus, W. R., and Anderson, J. L., 1977, Gneiss and migmatite of Archean age in the Precambrian basement of central Wisconsin: Geology, v. 5, p. 45–48.

Van Schmus, W. R., Medaris, L. G., Jr., and Banks, P. O., 1975, Geology and age of Wolf River batholith, Wisconsin: Geological Society of America Bulletin, v. 86, p. 907–914.

Van Schmus, W. R., Thurman, E. M., and Peterman, Z. E., 1975, Geology and Rb-Sr chronology of middle Precambrian rocks in eastern and central Wisconsin: Geological Society of America Bulletin, v. 86, p. 1255–1265.

Van Schmus, W. R., Woronick, R. E., and Egger, N. L., 1978, Geochronologic relationships in the Carney Lake Gneiss and other basement gneisses in Dickinson County, upper Michigan: Abstracts and Proceedings, 24th Institute on Lake Superior Geology, Milwaukee, Wisconsin, p. 39.

Weidman, S., 1907, The geology of north-central Wisconsin: Wisconsin Geological and Natural History Survey Bulletin 16, 697 p.

White, D. A., 1954, Stratigraphy and structure of the Mesabi range Minnesota: Minnesota Geological Survey Bulletin 38, 92 p.

MANUSCRIPT ACCEPTED BY THE SOCIETY MARCH 4, 1982

Geological Society of America
Memoir 160
1983

Tectono-sedimentary history of early Proterozoic rocks of the northern Great Lakes region

Grant M. Young
Department of Geology
University of Western Ontario
London, Ontario N6A 5B7
Canada

ABSTRACT

Paleocurrents, facies, and thickness variations suggest that volcanic and sedimentary rocks of the lower Huronian succession accumulated in an easterly-trending fault-bounded trough. The upper Huronian, here taken to include the glaciogenic Gowganda Formation and younger Huronian units, is much more widespread and records deposition under conditions of more general downwarping of the crust. In the Lake Superior region, rocks of the Chocolay Group, which includes tillites and extensive orthoquartzites, are considered to be part of this depositional phase. Paleocurrents in both the Huronian Supergroup and Chocolay Group are generally eastward-directed during these two depositional phases.

A strong deformational event next affected rocks of the Lake Huron region. This event is dated at about 2.1 Ga by contemporaneous intrusion of the Nipissing diabase. There is some evidence that much of the Huronian Supergroup was semilithified during this phase of deformation (McGregor phase). A similar tectonic episode may be recorded on the south shore of Lake Superior by intrusion of a porphyritic red granite about 2.0 Ga ago and by deformation and uplift of rocks of the Chocolay Group prior to deposition of the Menominee Group. The name Michigan phase is proposed for this event. In Minnesota, diabase dykes of similar age to the Nipissing diabase of Ontario were apparently intruded before deposition of the Menominee Group.

The next depositional episode is characterised, in the Lake Superior region, by deposition of basal conglomerate and orthoquartzite followed by iron-formation. Equivalent rocks are not known in the Lake Huron region which may have been weakly emergent in the aftermath of the McGregor phase. In the Lake Superior region, the final phase of recorded early Proterozoic sedimentary history is represented by a succession of carbonaceous mudstones followed in a coarsening-upward sequence by turbidites. This phase may be represented in the Lake Huron area by mudstones and proximal turbidites of the Whitewater Group, preserved uniquely in the Sudbury Basin. Turbidites of the Whitewater Group were transported westward; there is some evidence of similar transport directions in the corresponding rocks in the Lake Superior region, but significant amounts of clastic detritus were drawn from the north and south margins of the basin.

Subsequent deformation and low pressure, intermediate temperature metamorphism during the Penokean orogeny affected early Proterozoic rocks throughout the northern Great Lakes region. By analogy with other Proterozoic and Phanerozoic depositional basins, the sedimentary and tectonic history of early Proterozoic rocks of the northern Great Lakes area is tentatively interpreted in terms of the aulacogen model; a corollary of this interpretation is that the aulacogen opened into an ocean lying somewhere to the east, in the area now occupied by the Grenville province.

INTRODUCTION

The tectonic setting of the Huronian Supergroup (Fig. 1) has been the subject of some debate. It has been considered as a tectonically segmented platform at the edge of an Archean protocontinent (Frarey and Roscoe, 1970) and as a marginal paraliageosyncline in which subsidence was tectonically controlled (Young, 1971). Dietz and Holden (1966) compared the Huronian to the miogeosynclinal portion of a classical geosynclinal couplet; the eugeosyncline was thought to be represented by part of the Grenville province. Church (1971) proposed an origin as a simple trough like the Appalachian geosyncline at an early stage in its evolution. Parviainen (1973) and Card (1978a) considered the possiblity that the Huronian succession is bounded to the south by Archean rocks so that the depositional basin was an elongate, easterly-trending graben-style trough. The

southern margin of the Huronian is largely unknown because it is covered by lower Paleozoic rocks of the northeastern margin of the Michigan Basin and the waters of Lake Huron. Geophysical studies (O'Hara and Hinze, 1980) suggested, and drilling confirmed, the presence of gneisses to the south of the Huronian outcrop belt in Manitoulin Island. These gneisses have a Rb-Sr whole rock age of about 1.8 Ga, which was considered by Sims and others (1980) to be a metamorphic overprint (Van Schmus and others, 1975) on rocks of Archean age.

An additional problem has been the nature of the contact between the eastern part of the Southern Province (Huronian outcrop belt) and the Grenville province to the east (see Card, 1978a, p. 6 for a summary). The most recent opinion is that the Grenville Front is a very ancient tectonic

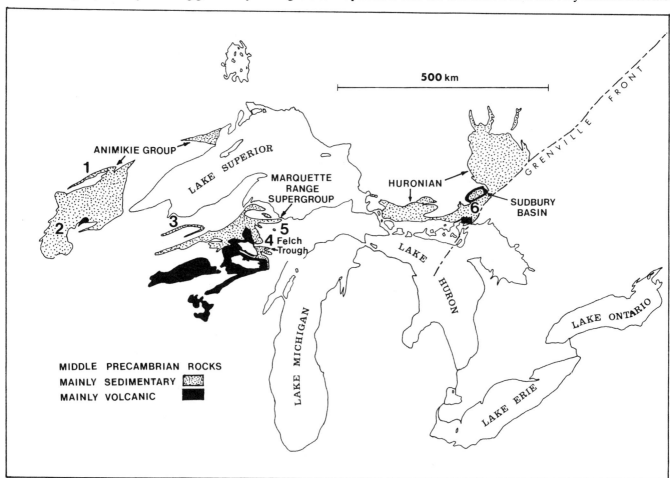

Figure 1. Location of major areas of early Proterozoic supracrustal rocks in the Great Lakes area. Numbers show the locations of stratigraphic sections in Figure 2. Small black rectangle below the numeral 6 shows approximate location of Figure 4.

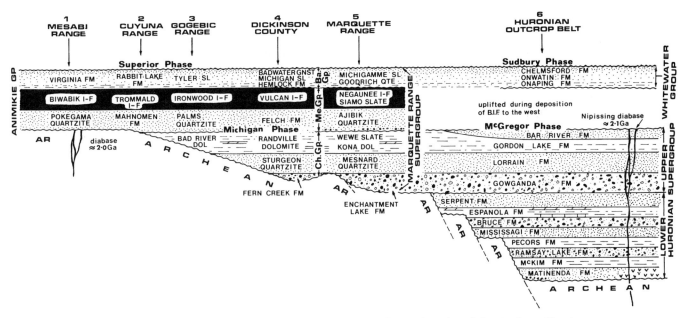

Figure 2. Schematic representation of early Proterozoic stratigraphy of the northern Great Lakes region. Numbers refer to locations shown in Figure 1. See text for discussion.

zone, extending back at least to early Proterozoic time. Card (1978a), Lumbers (1975), and Schwerdtner and Lumbers (1980) described the Grenville Front as coinciding with a paleogeographic hinge line separating the typical "miogeosynclinal" Huronian facies to the northwest from a deeper water facies to the southeast.

Regional interpretations have also been frustrated by a lack of agreement concerning stratigraphic relationships between the Huronian Supergroup and early Proterozoic rocks of the Lake Superior region. For a summary of earlier attempts at correlation, see Young (1966) and Church and Young (1970). It was suggested by these authors that most of the Huronian is older than most of the Marquette Range Supergroup (Fig. 2) of the Lake Superior region but that the more extensive upper part of the Huronian succession might be correlated with the lower part of the Lake Superior succession, the Chocolay Group. Largely on the basis of geochronological studies, these correlations have been considered by most subsequent workers to be incorrect (e.g., Morey, 1973; Roscoe, 1973; Van Schmus, 1976; Card, 1978a); and there has been a growing tendency to interpret the early Proterozoic rocks of the Lake Superior region as a separate entity.

Geologists in the United States have been much more ambitious than their Canadian counterparts in attempting to apply plate tectonic concepts to early Proterozoic rocks of the Great Lakes region. In recent years, the following interpretations have been made:

1. Sims (1976), Morey and Sims (1976), and Sims and others (1980) proposed that the Marquette Range Supergroup accumulated on sialic basement. They recognized two contrasting types of Archean terrane: to the southeast

an ancient (>3.0 Ga) granite-gneiss terrane, and to the northwest a younger greenstone-granite terrane. The boundary between these two terranes, the Great Lakes tectonic zone, was considered important in location of the depositional basin(s) in which the early Proterozoic rocks accumulated. Deposition and deformation of the Proterozoic rocks was attributed to greater mobility of the ancient gneissic crust.

2. Van Schmus (1976) suggested that the early Proterozoic rocks of the Lake Superior region represent a southward-facing miocline that was subsequently deformed and intruded by plutonic rocks as a result of development of a northward-dipping subduction zone. Discovery of possible ancient Archean basement gneisses in central Wisconsin later led Van Schmus and Anderson (1977) to suggest modification of the earlier interpretation in favor of that of Sims (1976) and Morey and Sims (1976).

3. Cambray (1977; 1978) interpreted volcanic and intrusive igneous rocks south of the main part of the Penokean fold belt on the south shore of Lake Superior as the result of continental collision following consumption of oceanic crust in a south-dipping subduction zone.

4. LaRue and Sloss (1980) proposed development of a series of separate depositional troughs in the Lake Superior region and suggested that early Proterozoic volcanic rocks to the south were juxtaposed against the more northerly, rift-related sedimentary basins by means of a south-dipping subduction zone.

The main purposes of this paper are to re-examine possible stratigraphic relations between the early Proterozoic successions of the Lake Huron and Lake Superior regions and to attempt an integrated summary of the depositional and tectonic history of this region. The depositional history during early Proterozoic time is here considered in terms of three major stages. Deformation occurred between stages 2 and 3 and after stage 3. These stages are similar to those described by Hoffman and others (1974) in connection with the Athapuscow aulacogen in the Northwest Territories of Canada.

STAGE 1: GRABEN STAGE

This stage is represented by the lower part of the Huronian succession; these rocks have a limited areal distribution, mainly in the southern part of the Huronian outcrop belt (Fig. 1; 3a). The stratigraphic succession of the Huronian is shown schematically in Figure 2. The Huronian Supergroup is divided into four groups but for simplicity is here considered as lower and upper Huronian. The lower Huronian includes middle Precambrian rocks older than the Gowganda Formation. It comprises up to 6,000 m of sedimentary and volcanic rocks but thicknesses vary greatly from place to place. Several formations are characterised by rapid thickness changes to the south, across synsedimentary faults. An important advance in the last decade has been the discovery of widespread mafic and felsic volcanic rocks in the basal part of the Huronian succession (Robertson, 1973; Sawiuk, 1977; Card, 1978a). These have been described in detail by Card (1978a) who interpreted them as fissure eruptions controlled by deeply penetrating marginal block faults. On the basis of chemical studies, Sawiuk (1977) compared the Thessalon volcanics of the basal Huronian with those of the Red Sea-Afar system. Layered anorthosite-gabbro intrusions are common at the base of the Huronian succession in the eastern part of the fold belt. These were interpreted by Card (1978a) as being comagmatic with the volcanic rocks which were extruded about 2400 Ma ago.

The Creighton and Murray plutons are mainly quartz monzonite; they were considered by Card (1978a) to be late-stage epizonal magmatic intrusions related to the Huronian vulcanicity, but the radiometric ages suggest that they may be considerably younger. Dutch (1979) considered the Creighton granite to be a forcefully intruded diapir. Basement granites of the Superior province to the north have yielded minimum radiometric ages of about 2500 Ma (Van Schmus, 1965). Rb-Sr isochrons from the basal Huronian volcanics have given radiometric dates that range from 2005 ± 125 Ma to 2495 ± 145 Ma (Knight, 1967). An Rb-Sr "scatterchron" (age of 2230 ± 20 Ma was obtained from the Murray Granite (Gibbins and others, 1972). According to Fairbairn and others (1960; 1965), the Rb-Sr whole rock age of both the Murray and Creighton granites is about 2.14 Ga. All of these dates are somewhat suspect.

There is a general southerly thickening of many Huronian formations, particularly the sandstones and mudstones. Some glaciogenic(?) conglomerates such as the Ramsay Lake and Bruce Formations decrease in thickness towards the southeast, possibly reflecting the limits of glacial ice. Much thickening is related to down-faulting along marginal growth faults such as the Murray fault. Additional evidence of contemporaneous faulting is provided by resedimented breccias and conglomerates in the vicinity of the Murray fault (Long, 1977).

More obvious north-to-south thickening and facies changes have tended to overshadow some equally important changes that occur from west to east. There is a general increase in thickness of many formations towards the east. Fluvial deposits of the Matinenda Formation are replaced in the east by volcanic rocks. The McKim Formation shows *facies and thickness changes* in that direction, as also does the Pecors Formation (Card, 1978a). Fluvial sandstones of the Mississagi Formation (Long, 1978) exhibit some of the most spectacular thickness changes; Card (1978a, p. 120) estimated a west-to-east rate of thickening of the unit at about 50 m per kilometer.

The Matinenda and McKim Formations in the basal part of the Huronian succession record initial subsidence of the trough with concomitant vulcanism that is especially marked in the east. The presence of coarser facies in the turbidites of the McKim Formation towards the east may reflect greater uplift and faulting (Innes and Colvine, 1979) along the basin margins in that area. The increased thickness of many sedimentary formations also suggests significant basinal subsidence in the east.

As pointed out by Roscoe (1969), the succeeding six formations of the lower Huronian comprise two cycles, in part tectonically controlled, but possibly also reflecting responses to glacial activity. The cycles begin with glaciogenic conglomerates and associated rocks, overlain by finer grained, deeper water deposits that grade upward through deltaic silty mudstones to thick sandstones that are dominantly fluviatile. The Espanola Formation (Fig. 2), which is the middle unit of the second cycle, is largely composed of carbonate (Young, 1973a), containing some stromatolites (Hofmann and others, 1980) and some evidence of tidal influence (Young and others, 1977).

Paleocurrents in rocks deposited during Stage 1 indicate dominant transport to the east and southeast, especially in thick sandstones interpreted as fluvial sediments (Long, 1978). More complex paleocurrent patterns at the eastern end of the outcrop belt may be related to contemporaneous northeast- and northwest-trending faults (Innes and Colvine, 1979). The paleocurrent data, together with thickness and facies changes, indicate that deposition took place in a fault-bounded trough that trended approximately east-west and underwent greater subsidence at the eastern end (Fig. 3a). The dominantly longitudinal fluvial transport suggests that the lower Huronian rocks did not

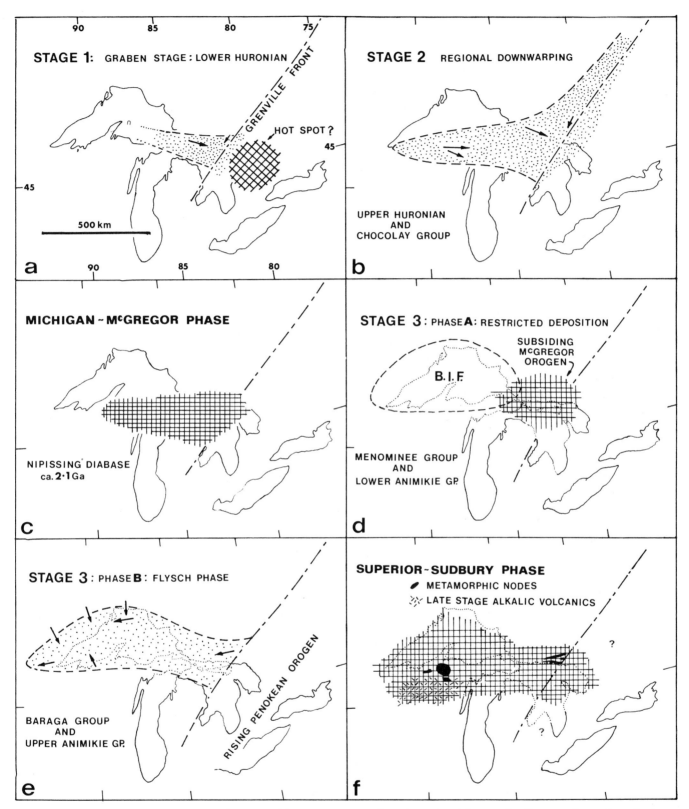

Figure 3. Some major depositional and tectonic events involving early Proterozoic rocks of the northern Great Lakes region. See text for discussion.

accumulate on the southern margin of an easterly-trending continental margin (Dietz and Holden, 1966) but rather that a confining boundary existed to the south. Lumbers (1975), Card (1978a) and Schwerdtner and Lumbers (1980) suggested that the Grenville Front may coincide approximately with a major facies change in some Huronian formations; deeper water deposits were considered to lie to the east in the Grenville province.

Rocks of Stage 1 are not represented in the Lake Superior region (Fig. 3a).

STAGE 2: REGIONAL DOWNWARPING

Lake Huron Region

In the Huronian succession, Stage 2 (Fig. 3b) is represented by four formations with a total thickness of about 5,000 m. These units are generally conformable on the lower Huronian in the southern part of the outcrop belt; but in more northerly regions, there is local evidence of unconformable relations, possibly related to tilting of fault blocks or early folding. The Gowganda Formation, the lowest unit of the upper Huronian has long been considered to be glaciogenic (for a recent summary see Young, 1981). To the north the upper Huronian formations overstep the lower Huronian and lie directly on Archean basement rocks. Wider distribution and preservation suggest a more general subsidence permitting first accumulation of Proterozoic sediments in the region north of the initial graben that formed during Stage 1 (Fig. 3b). Glacial rocks of comparable age may have been widespread in North America (Young, 1970; 1973b), permitting approximate time correlation of widely separate regions.

The succeeding formations of the upper Huronian contain evidence suggesting a dramatic climatic amelioration. Aluminous minerals such as kaolinite and diaspore and metamorphic derivatives (Church, 1967) are common in the Lorrain Formation (Fig. 2). These were attributed by Chandler and others (1969) to contemporaneous intense weathering of feldspar. Wood (1973) documented the occurrence of gypsum and anhydrite in the Gordon Lake Formation and reported hematitic ooliths.

Lake Superior Region

There is no definitive evidence in these westerly regions, of deposition during lower Huronian times. It was suggested by Young (1966) and Church and Young (1970) that a stratigraphic correlation is possible between the rocks of the Chocolay Group (Fig. 2) and the upper Huronian succession. This correlation has subsequently been denied on the basis of geochronological data but is reaffirmed here. The strongest evidence for the correlation is the presence of glaciogenic rocks at the base of the Chocolay Group

in a number of locations in northern Michigan. These include the Reany Creek Formation (Puffet, 1969), the Fern Creek Formation (Pettijohn, 1943), and the Enchantment Lake Formation (Gair, 1975). The glaciogenic rocks are in turn overlain by extensive orthoquartzites, some of which are aluminous (Gair, 1975) like their putative correlative, the Lorrain Formation.

The topmost unit of the Chocolay Group, the Kona dolomite and equivalents, is fine-grained, chert-like and similar in many respects to the Gordon Lake Formation of the upper Huronian. Recently, Hofmann and others (1980) described dolostone with fenestral fabric from the Gordon Lake Formation, providing limited evidence of benthic microbial action within the dominantly fine clastic depositional environments of the Gordon Lake Formation. The Kona dolomite, in contrast, contains abundant stromatolites (Gair, 1975); but its general appearance and cherty nature are reminiscent of the Gordon Lake Formation. The differences between these units could be the result of greater clastic influx into the depositional site of the Gordon Lake Formation.

Correlation of Rocks of Stage 2

The lithological similarities mentioned above have been discounted because of results of geochronological studies. For example, in Minnesota diabase dykes dated at about 2.0 Ga to 2.2 Ga (Hanson and Malhotra, 1971) were considered to have been intruded prior to deposition of rocks of the Animikie Group (Fig. 2). In the area where the dykes occur, however, there are no representatives of the Chocolay Group so that the relationship between these diabases and the Chocolay Group is unknown. In northern Michigan early Proterozoic diabase dykes cut rocks as young as the Michigamme Formation (Sims and others, 1980). Clearly there are early Proterozoic diabases of various ages in the Lake Superior region, but the older ones (ca. 2.0 Ga) have not been shown to be older than the rocks of the Chocolay Group; they may, however, be older than deposition of the Menominee and Baraga Groups (Fig. 2).

Another line of evidence that has been used to suggest that all of the Marquette Range Supergroup is younger than about 2.0 Ga concerns a porphyritic red granite in basement rocks just north of the Felch Trough in Michigan (Fig. 1). Because of the presence of foliation and lineation in the granite and their absence in rocks of the Chocolay Group a few kilometers distant, James and others (1961) considered the red porphyritic granite as "definitely pre-Animikie." Contacts between the porphpyritic granite and rocks of the surrounding Archean Dickinson Group are poorly exposed.

The porphyritic red granite has yielded an Rb-Sr age of about 2.0 Ga (Van Schmus, 1976), and U-Pb analyses of zircons define an age of about 2.1 Ga (Banks and Van

Schmus, 1972). Van Schmus (1976) used these ages to suggest that the rocks of the Chocolay Group are younger than about 2.0 Ga. Because the geological relationships in this area are not clear, it is difficult to evaluate the significance of these radiometric ages. It is, however, instructive to compare recent studies of a very similar foliated porphyritic granite in the lower part of the Huronian near Sudbury (Dutch, 1979). A particularly close analogy is possible with the Creighton pluton. According to Fairbairn and others (1960; 1965), both the Creighton and Murray granites have Rb-Sr whole rock ages of about 2.14 Ga, close to those obtained from the porphyritic red granite in Michigan. Recently, Frarey and others (1982) published a U-Pb zircon date of 2333 + 33/-22 Ma for the Creighton granite and interpreted this date as a minimum age for deposition of the entire Huronian Supergroup. Card (1978a), however, suggested that the Creighton granite might be genetically related to the Basal Huronian volcanics. Dutch (1979) mapped a "wrap around" foliation in the Creighton granite and contrasted it to the relatively weak development of penetrative fabric in the surrounding rocks. He concluded that the fabric in the granite was the result of forceful emplacement as a diapir although some of these conclusions were questioned by Card (1979). Close similarities are found in the red porphyritic granite of Michigan where James and others (1961) noted, and their map clearly illustrates, that the attitudes of foliation in the granite "appear to define a domelike structure." By analogy with the Creighton granite, which is post-Huronian yet is more strongly foliated than some of its host rocks, the red porphyritic granite *could be interpreted as a post-Chocolay Group diapir in which the strong "wrap around" foliation is related to forceful intrusion.* Thus the Chocolay Group could be older than the 2.0 Ga age of the granite.

Similar radiometric ages from basement gneisses in the vicinity of the Felch Trough were interpreted by Van Schmus (1976) as representing a metamorphic episode prior to deposition of the Chocolay Group. Aldrich and others (1965), however, reported an age of about 2.0 Ga (Rb-Sr from metamorphic muscovite) from the Sturgeon quartzite in the Felch Trough so that the 2.0 Ga to 2.1 Ga event may be later than deposition of the Chocolay Group and affected both basement and supracrustal rocks.

In many places, particularly in the eastern part of the Lake Superior outcrop belt, there is evidence of an unconformity between the Ajibik quartzite and underlying rocks of the Chocolay Group. The magnitude of this unconformity is uncertain; but according to Gair (1975), large fragments of folded and veined Kona dolomite occur in the basal conglomerate of the Ajibik Formation, which cuts across the underlying early Proterozoic formations to come to rest on Archean basement. These results suggest that there was a period of folding and uplift preceding deposition of the Menominee Group. It is suggested that these

events may correspond approximately to the folding that, in the Lake Huron region, affected the entire Huronian succession and preceded and accompanied intrusion of the Nipissing diabase about 2.1 Ga ago. If this interpretation is correct, then Stage 2 was terminated by a period of deformation that was more strongly expressed in the eastern part of the Lake Superior region and in the Lake Huron outcrop belt. This deformation resulted in development of a major unconformity in the Huronian outcrop belt where the Superior-type iron-formations are missing and Stage 3 began with deposition of the Whitewater Group of the Sudbury Basin (Figs. 1, 2).

STAGE 3: RESTRICTED DEPOSITION (A) AND FLYSCH PHASE (B)

Deposits of the period of restricted deposition (A) are not preserved in the Huronian outcrop belt which may have been weakly positive in the aftermath of the McGregor phase of deformation (Fig. 3c). Sedimentation recommenced in the Lake Huron region with unconformable deposition of the Whitewater Group on the folded Huronian.

Restricted Depositional Phase (3A)

Stage 3 was ushered in by basal conglomerate and quartzite of the Ajibik Formation. This was followed by widespread deposition of iron-formation in the Lake Superior region. Detailed consideration of the depositional environments of the iron-formations is beyond the scope of this paper. Gair (1975) noted that there was local influx of clastic debris into the dominantly chemical environment in which the iron-formation accumulated. Morey (1972, p. 210) presented arguments against a volcanic origin for the iron and tentatively suggested that concentration of iron by evaporation of sea water may have been important. If the Huronian region were slightly emergent, the depositional basin in which the iron-formations precipitated could have been at least partially cut off from normal circulation during this phase (Fig. 3d). There is some evidence of climatic change following the Gowganda glaciation so that a warm climatic regime with possible evaporitic conditions could have led to further concentration of iron in already iron-rich water. The basin would have been generally starved of clastic material prior to the arrival of mud and sand from the rising orogen to the east (Fig. 3e).

Flysch Phase (3B)

The final early Proterozoic depositional event recorded in the Lake Superior region is comparable to the flysch phase of the Athapuscow aulacogen. It is represented in the Lake Superior region by, for example, the Rove, Virginia, and Rabbit Lake Formations, the Michigamme

Slate and other units, all of which show a gradational relationship with underlying iron-formations. These units are generally of deeper water aspect, suggesting that basin subsidence outstripped sedimentation rate (Morey, 1972). A wide variety of rock types are included in these formations (see Morey, 1973 for a detailed stratigraphic and mineralogical analysis). In general, the succession begins with mudstones of deep-water aspect. These grade upward into coarser grained greywackes deposited largely by the action of turbidity currents. Many of the units are black, reflecting the presence of carbon; carbonate concretions are common.

The provenance of this prograding flysch sequence is critical to the interpretation presented here. Morey (1973) proposed a major northerly source for the formations of Phase 3B on the north shore of Lake Superior. Minor southerly sources were also documented by Morey (ibid.), and LaRue and Sloss (1980) proposed the existence of a southerly source terrane in the vicinity of the Gogebic Range. Morey (1972, p. 223), from a study of petrography and paleocurrents, suggested that the main source of coarse clastic material in the Rove Formation on the north shore of Lake Superior was a basement terrane to the north. Paleocurrent diagrams (Morey, 1972, p. 222) based on cross beds and ripple marks show a prominent southwesterly component, and Morey (ibid., p. 223) noted that the source of the fine material remains a problem. In a more regional study, Morey (1973, his Fig. 10) also documented some westerly transport in the Lake Superior region. Keighen and others (1972), on the basis of cross-bedding studies, suggested southerly transport of the material making up the Thomson Formation in east-central Minnesota; they also documented both easterly and westerly flow based on flute casts and a dominantly east-west orientation of groove casts.

Lake Huron Region

It is suggested that correlatives of the flysch phase (3B) of the Lake Superior region may be uniquely preserved in the Whitewater Group of the Sudbury Basin (Church and Young, 1972). There are many lithological similarities between the Whitewater Group and Phase 3B rocks of more westerly regions. These include the presence of carbon, carbonate concretions, turbidites, and the generally coarsening-upward nature of the succession.

The Sudbury Basin has been attributed by a number of authors to meteorite impact, but the evidence is equivocal (Guy-Bray, 1972; Card and Hutchinson, 1972; Fleet, 1979, 1980). Central to these arguments is the nature and origin of the basal clastic unit in the Sudbury Basin, the Onaping Formation. Some authors (e.g., Peredery, 1972) have claimed that these rocks are a fall-back breccia associated with meteorite impact, but Stevenson (1972) maintained that they are ash flow deposits or ignimbrites.

The overlying rocks include the Onwatin Slate and associated thin basal limestones with Pb-Zn-Cu ores (Card and Hutchinson, 1972). These are followed in a generally coarsening-upward sequence by a thick, immature sandstone succession interpreted as a series of proximal turbidites. Studies by Rousell (1972) and Cantin and Walker (1972) have shown that the turbidites were not deposited centripetally in a circular depression as the astrobleme theory might demand but rather that there was a consistent west-southwesterly flow of turbidity currents across the basin. The turbidites of the Chelmsford Formation retain their proximal character throughout the length of the Sudbury Basin. This was interpreted by Cantin and Walker (1972) as indicating the existence of a WSW-ENE-trending elongate depositional basin in the present site of the Sudbury Basin. The persistent coarse nature of the sediments was explained as being due to lateral sediment supply from the northwest and southeast, with superimposed axial flow to the WSW. These results could also be interpreted as showing that the sedimentary rocks of the Whitewater Group are merely the small preserved remnants of a formerly extensive westerly-transported flysch succession. The lack of a facies change within the Sudbury Basin could be because the preserved part of the succession is a small part of a regionally extensive flysch apron. Preservation in the Sudbury Basin may be due to downfolding prior to intrusion of the Sudbury Irruptive (Morris, 1980), which later provided a protective resistant upstanding rim. Subsequent uplift of the region to the south, in part along the ancient bounding faults of the main depositional basin, could have led to removal of these younger rocks.

Radiometric dating of the Sudbury Irruptive has led to somewhat conflicting results, but the main period of intrusion of the norite is generally taken to be about 1850 Ma-1950 Ma (Gibbons and others, 1972; Fairbairn and others, 1969; Card, 1978b).

OROGENIC HISTORY

Terminology

Much of the current confusion concerning the early Proterozoic evolution of the Great Lakes region centres on the use of the term "Penokean orogeny." In some cases, the term has been used in a restricted sense for a single "orogenic phase." For example, Dutch (1979), following Brocoum and Dalziel (1974), considered the Penokean orogeny to be represented by the deformational event that postdated the intrusion of the Sudbury Irruptive. In other cases, it has been used in the sense of an "orogenic cycle" to cover all the events involved in the evolution of a structural province (Church and Young, 1972, Table II; Cannon, 1973, p. 252) or at least the polyphase deformation (Church and Young, 1972, Table I; Gibbins and McNutt, 1975, p. 1998).

This fundamental difference in the use of the term "orogeny" has been compounded by a lack of agreement concerning the age and correlation of deformational events recognised throughout the Great Lakes region. Van Schmus (1965, p. 771) concluded that folding of the Huronian of Ontario may not be true "Penokean" folding since it may be 300 Ma older than the main activity of the Penokean folding in the area south of Lake Superior, which Goldich and others (1961) placed in the interval 1800 to 1700 Ma. Church (1966) took the opposite view, suggesting that the main phase of Penokian folding took place at least as early as 2.1 Ga ago, and rejected the 1.8 Ga to 1.7 Ga K-Ar dates as survival ages unrepresentative of the Penokean orogeny. Later, Church (1968) emphasized that the Penokean was a polyphase event involving three phases of deformation. The earlier two phases were considered to be older than 1900 Ma, and the third coincident with the Hudsonian orogeny at 1.7 Ga ago. The second phase, of unknown radiometric age, was considered to postdate the formation of the enigmatic Sudbury breccia and shatter cones that are widespread throughout the Huronian. Church (1968) was also undecided as to whether the early two phases of deformation should be considered as entirely separate orogenic events or might be initial phases of crustal breakdown leading to the extensive cycle of Hudsonian magmatic activity.

On the basis of Rb-Sr (Wetherill and others, 1965) and U-Pb age data (Banks and Van Schmus, 1971; Cannon, 1973) revised their estimate of the minimum age of the main phase of Penokean deformation in Michigan to a value of 1900 Ma. Cannon (1973, p. 270) emphasized, however, that "this is the date for the final events of the Penokean orogeny and that a long sequence of deformation occurred prior to that time." Van Schmus and others (1975) proposed that granitic rocks as young as 1800 Ma should also be considered part of the Penokean orogeny. Gibbins and McNutt (1975) suggested that the Penokean orogeny in Ontario was represented by the main phase cleavage in the Huronian of the Sudbury region and was as young as 1800 Ma-1600 Ma. This phase was also chosen by Brocoum and Dalziel (1974), Card (1978a), and Dutch (1979), all of whom considered the main phase cleavage, both inside and outside the Sudbury Basin, to postdate emplacement of the Sudbury Irruptive and therefore to be younger than 1850 Ma (Krogh and Davis, 1974) or about 2.0 Ga (Rb-Sr data of Gibbins and McNutt, 1975). On the other hand, Van Schmus (1976) proposed that the main phase cleavage in the Huronian rocks of the Sudbury area predated formation of the Sudbury Basin and was therefore at least as old as 1.9 to 1.8 Ga. More recently Card (1978b; 1979) has proposed that the main postbreccia deformation is as young as 1.8 to 1.7 Ga, referring the Penokean orogeny to some undefined event involving deformation and metamorphism about 1.9 Ga ago.

In this paper, the term Penokean orogeny is used to include the main phases of deformation and attendant thermal and igneous events that affected the early Proterozoic rocks of the Great Lakes region between about 1.8 and 2.1 Ga. ago. The first phase of deformation in the Lake Huron region is here called the "McGregor phase"; the name is taken from McGregor Bay in the northern part of Lake Huron where a major anticline is attributed to the early deformational episode. This episode is correlated with folding, uplift, and erosion in northern Michigan between deposition of the Chocolay and Menominee Groups, here ascribed to the "Michigan phase." The main phase of the Penokean orogeny in the Lake Superior region was considered to have terminated about 1.82 to 1.9 Ga ago (Cannon, 1973; Van Schmus, 1976; 1980). This phase is called the "Superior phase" and is considered approximately contemporaneous with the deformation and metamorphism of the Huronian, and the Whitewater Group in the Sudbury region. This phase, in Ontario, is called the "Sudbury phase." These names are shown on Figure 2 and are listed in Table 1.

In northern Michigan the Archean basement rocks underwent little compression during the Penokean orogeny but were affected by block faulting. This is in contrast to central Wisconsin where Archean rocks may have been subjected to isoclinal Penokean folding (Maas and others, 1980). Results of U-Pb dating of zircon by Van Schmus (1980) from early Proterozoic rocks of Wisconsin led to the following interpretations. A phase of mafic-to-felsic vulcanism dated at 1859 ± 20 Ma was taken by Van Schmus (1980) as the onset of the main pulse of the Penokean orogeny. The main peak of orogenic activity was considered to be contemporaneous with tonalitic and granitic plutons dated at about 1850 to 1820 Ma. It was previously suggested by Cannon (1973) and Van Schmus (1976) that the Penokean orogeny in northern Michigan and Minnesota terminated about 1.85 to 1.90 Ga ago.

A younger granite-rhyolite suite in central Wisconsin has yielded distinctly younger radiometric ages of about 1760 ± 10 Ma. These rocks may possibly be a late-stage episode of the Penokean orogeny (Smith, 1978). Anderson and others (1975) proposed that the granite-rhyolite suite was formed by melting of tectonically thickened crust during the last stages of the Penokean orogeny. These rhyolites are overlain by a succession of sandstones such as the Baraboo quartzite; these clastic rocks appear to have been derived mainly from the north (Dott and Dalziel, 1972).

Lake Superior Region

Most of the early Proterozoic rocks of the north shore of Lake Superior region are relatively mildly deformed. More intense deformation is evidenced in eastern Minnesota, Wisconsin, and northern Michigan. In the Michigan

G. M. Young

TABLE 1. POSSIBLE SEQUENCE OF EARLY PROTEROZOIC EVENTS IN THE NORTHERN GREAT LAKES REGION

		Event	Lake Huron Area	Lake Superior Area	Age (Ma)
EARLY PROTEROZOIC	PENOKEAN OROGENY	Granite intrusion, Late rhyolitic vulcanism and latest phases of deformation	Grenville Front and Cutler plutons	Volcanic rocks of southern Wisconsin	1.7-1.8
			?Sudbury Irruptive?*	-	1.84?
		Late orogenic phase, intrusion and metamorphism	*SUDBURY PHASE*	*SUPERIOR PHASE* Granites and tonalites, southern Wisconsin	Culmination at 1.82-1.90
			?Sudbury Irruptive, shatter cones and Sudbury breccia?		
		Stage 3B (flysch phase)	Whitewater Group (NE provenance)	Baraga Group, Upper Animikie Group (N, S, and E provenance)	
		Stage 3A (quartzite and BIF)	?Sudbury breccia and* shatter cones?	Menominee Group	
		Early orogenic phase and metamorphism	Nipissing diabase *McGREGOR PHASE*	unnamed diabase *MICHIGAN PHASE*	2.1
		Stage 2 (regional downwarping)	Upper Huronian (NW provenance)	Chocolay Group (NW provenance)	2.2-2.3
		Stage 1 (Graben formation)	Lower Huronian (NW provenance)	-	2.3-2.4?

UNCONFORMITY

ARCHEAN ROCKS

*Favored interpretation

area Cannon (1973) considered the primary, generally easterly-oriented deformation structures to be due to gravity sliding. Later folds were interpreted by Cannon (ibid.) as due to accommodation of block faulting of basement rocks. Maas and others (1980) proposed that there was more basement involvement in Penokean deformation in central Wisconsin.

Metamorphism in the area south of Lake Superior is in the form of nodes (Fig. 3f), as pointed out by James (1955). Some of the nodes are not obviously associated with plutonic igneous activity, but others are cored by granitic plutons about 1900 Ma old (Aldrich and others, 1965; Banks and Van Schmus, 1971; 1972). Metamorphism is of low pressure, low-to-intermediate temperature type (Van Schmus, 1976). The nodal distribution is interpreted as indicating local thermal highs rather than being due to later folding of originally planar isothermal surfaces. The basement underwent little compression in the north near the south shore of Lake Superior, but farther south may have been involved in Penokean folding (Maas and others, 1980).

Lake Huron Region

Deformation and metamorphism in the region north of Lake Huron have been described recently by Card (1978b). Early deformation (~2.1 Ga ago) was followed by intrusion of diabase of the Nipissing suite (Church, 1966). There is some evidence that early deformation of the Huronian took place when the sediments were only partially lithified. For example, there are abundant clastic dykes in the Huronian succession (Young, 1972; Chandler, 1973; Card, 1976), some of which cut across fold axes, particularly in the Espanola Formation (Young, 1973a). Mapping in the southern part of the Huronian outcrop belt shows highly irregular and chaotic folding (Fig. 4) and peculiar irregular fault contacts between the Espanola and Serpent Formations. Similar-structures were noted by Card (1978a, p. 190). Some of these contacts suggest megaslumps and clastic intrusions on a scale of kilometers (Fig. 4). Large blocks derived from the Mississagi and Serpent Formations at the west end of the area shown in Figure 4 are probably fragments ripped from adjacent formations during early

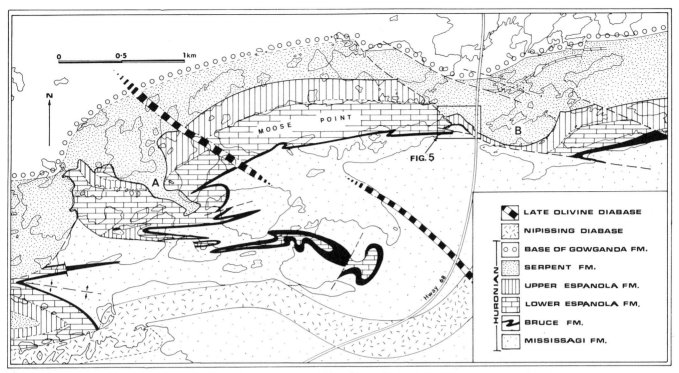

Figure 4. Geologic map of an area near the southern limit of the Huronian outcrop belt to illustrate chaotic folding, large-scale clastic intrusion (A) and slump(?) structure (B). These structures suggest that the sediment was only partially lithified at the time of their formation.

tectonic(?) activity. Detailed mapping (Fig. 5) shows that some formational contacts are ragged and irregular. Fragments of various formations up to tens of meters long are incorporated in the contact zones. These peculiar boundaries suggest tectonic/slump movements with partial flowage and partial fragmentation of the Huronian sediments along early fault contacts. The precise nature and timing of these events are not known, but the presence of Nipissing-type diabase along one of the early fault contacts (Fig. 5) indicates an early (pre-Nipissing diabase) origin.

Large areas of the upper Huronian and smaller areas of lower Huronian rocks lying north of the main fold belt are little deformed or metamorphosed; but in the southern and eastern parts of the outcrop belt, rocks of the Huronian Supergroup were affected by low-pressure-intermediate temperature type of metamorphism (Card, 1978b). As in the Lake Superior region, the metamorphism is in the form of thermal nodes; isograds cut across early fold structures. Card (1978b) stated that metamorphic rocks in the eastern Southern Province yielded Rb/Sr and K/Ar ages in the range of 1.6 to 1.9 Ga but also that the entire province was affected by regional metamorphism and deformation approximately 1900 Ma ago during the Penokean orogeny. Similar old ages (Rb-Sr isochrons) were obtained by Fairbairn and others (1969). Thus, the nature, style, and possibly the age of the main metamorphism are similar in both the Lake Superior and Lake Huron regions. In the Lake Huron area, these events affected the entire Huronian suc-

cession. The first phase of folding (McGregor phase) took place before deposition of the Whitewater Group (Card, 1978b; Morris, 1980). In the Lake Superior region, the first deformational event was generally less intense but was more strongly expressed in the east where a widespread unconformity separates the Ajibik quartzite and equivalent rocks from the underlying Chocolay Group (Fig. 2).

Card (1978a, p. 204–205) proposed a scheme in which early compression and low-grade metamorphism were followed by intrusion of the Nipissing diabase (~2.1 Ga ago) and formation of the Sudbury breccia and succeeding major deformation and metamorphism (Penokean?) affecting the Huronian, the Whitewater Group, and the Sudbury Irruptive. These events were followed by intrusion of the Grenville Front granitic plutons (1.6 to 1.7 Ga ago) and a succession of less important thermal and tectonic events possibly ranging in age up to the Grenville orogeny about 1.0 Ga ago. This scheme is similar in some respects to that adopted here but differs from that proposed by Card (1978b; 1979) in which the Penokean orogeny (~1.9 Ga ago) is considered to predate formation of the Sudbury breccias, the Whitewater Group, and the Nickel Irruptive.

The timing of intrusion of the Sudbury Irruptive is not known with certainty. If the 1.84 Ga date (Fairbairn and others, 1969; Krogh and Davis, 1974; Van Schmus, 1965) is accepted, then it must postdate the metamorphic culmination of the Penokean orogeny (1.85 to 1.90 Ga). If the 2.0 Ga age suggested by Gibbins and McNutt (1972; 1975) is

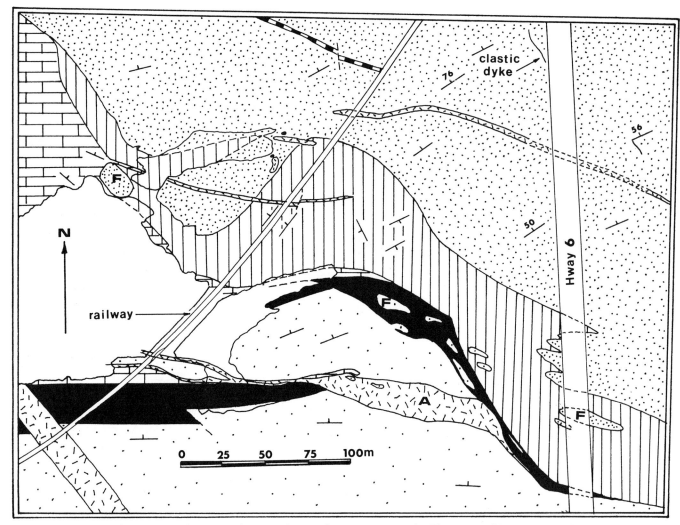

Figure 5. Detailed geologic map of part of the area shown in Figure 4 to illustrate ragged formational contacts due to early, preconsolidation movements. Note large fragments (F); diabase dyke (A) was intruded along an early fault. Symbols as in Figure 4.

correct, however, then its intrusion is older and preceded the deformation and metamorphism associated with the main phase of the Penokean orogeny as suggested by Brocoum and Dalziel (1974) and Fleet (1979; 1980) and inferred by Card (1978a). This interpretation is consistent with the age of the Sudbury breccias which are later than the first (McGregor) phase of deformation of the Huronian but were affected by the dominant cleavage here ascribed to the Superior phase of the Penokean orogeny. Some of these possibilities are shown in Table 1.

Brocoum and Dalziel (1974) proposed a single deformational and metamorphic event in the Sudbury region. They ascribed this event to the Penokean orogeny. There is clear evidence in many parts of the Huronian outcrop belt (Church and Young, 1972; Card, 1978b) for earlier and later deformational and metamorphic events. The suggestion of Brocoum and Dalziel (ibid.) that both the White-

water Group and the Huronian Supergroup were affected by the Penokean orogeny cannot be discounted, however, particularly if the older age for the Irruptive proves to be correct. Even accepting the 1.84 Ga age for the Irruptive, the Whitewater Group could still have been folded during the Penokean orogeny, for some paleomagnetic evidence suggests that the Whitewater Group was folded prior to intrusion of the Irruptive (Morris, 1980). Thus, there is still considerable uncertainty concerning the timing of specific intrusive and tectonic events in the Sudbury region. It is not known, for example, whether the deformation phase that folded the Whitewater Group inside the Sudbury Basin also produced the main post-Sudbury breccia cleavage in Huronian rocks outside the basin or is equivalent to a younger strain slip cleavage in the Huronian. Were the rocks within the basin deformed prior to intrusion of the Sudbury Irruptive as suggested recently by Morris (1980), and is later deformation of the Irruptive equivalent to the late strain slip cleavage outside the basin? Some possible time relationships are suggested in Table 1.

DISCUSSION

The above brief summary of the depositional, deformational, and metamorphic history of the early Proterozoic rocks of the Great Lakes region suggests that the Marquette Range and Huronian Supergroups were deposited in a complex, easterly-trending, fault-bounded ensialic trough or aulacogen. The depocenter was initially located in the Lake Huron area; but as fault activity moved toward the craton, similar troughs formed to the west. A more general subsidence later led to amalgamation of these troughs. A major paleocurrent reversal from dominantly easterly—along the length of the main trough—to westerly, may signify orogenic activity, possibly related to closure of an ocean somewhere to the east of the Huronian outcrop belt. The following points are relevant to the interpretation of this region as an aulogogen:

1. Initial normal faulting played an important role in sedimentation and vulcanicity, particularly in the eastern, oceanward(?) portion of the depositional basin.

2. Paleocurrent studies suggest initial easterly fluvial transport along the length of the basin, rather than normal to the basin axis as might be expected along a continental margin. Frarey and Roscoe (1970) and Pettijohn (1970) pointed out that finer grained facies equivalents of the thick Huronian sandstones are unknown in the Huronian outcrop belt. Lumbers (1975) and Card (1978a) noted significant facies changes in some Huronian formations in the vicinity of the Grenville Front, suggesting that finer grained equivalents of these formations were transported to the east into the area now occupied by the Grenville province. It is suggested in Figure 6 that the ocean to which the aulacogen(?) was related lay in the Grenville province; but subsequent deformation, uplift, and metamorphism preclude close definition. Schwerdtner and Lumbers (1980) delineated a large zone of early Proterozoic metasedimentary rocks extending up to 200 km southeast of the Grenville Front in the area southeast of the Huronian outcrop belt. They inferred (p. 167) that these rocks were regionally metamorphosed "between 2.16 and 1.8 Ga ago" and may be a remnant of the proposed geosynclinal zone.

3. The various stages in the sedimentary and tectonic evolution of a typical aulacogen are preserved. Formation of an initial graben was followed by a period of more regional subsidence which produced a sub-Gowganda unconformity in some areas. This phase completed the depositional history of the Huronian Supergroup which was then folded, faulted, metamorphosed, intruded by the Nipissing diabase about 2.1 Ga ago, and uplifted to produce a large semibarred(?) basin in the Lake Superior region. This basin

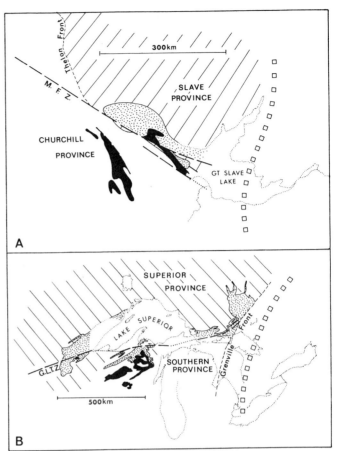

Figure 6. Comparison of the regional setting of the Athapuscow aulacogen in the Northwest Territories (after Hoffman, 1980) and the early Proterozoic rocks of the Great Lakes region. Cross-hatching represents Archean rocks; dotted ornament is early Proterozoic supracrustal rocks deposited in or near the aulacogen; black indicates syn-to-post-orogenic early Proterozoic supracrustal rocks. Note presence of major fault zones in both areas: MFZ is McDonald Fault Zone; GLTZ is Great Lakes Tectonic Zone (after Sims and others, 1980). Lines of squares show the approximate locations of hypothetical ocean basins. Note that the Athapuscow aulacogen is shown as a mirror image.

received basal conglomerates and quartz sand, followed by deposition in a starved basin, of the Superior-type iron-formations. A rising orogen in the east later contributed westerly-transported debris during the flysch phase, although much detritus was also drawn from the basin margins to the north and south. The major reversal of paleocurrent directions is one of the hallmarks of a cycle of ocean opening and closure (Hoffman, 1969; Hoffman and others, 1974). The flysch phase is marked by deposition of a coarsening upward succession from shales to turbidites, the proximal facies of which is uniquely preserved in the Sudbury Basin (Church and Young, 1972, p. 15).

Deposits comparable to the molasse and postorogenic Et-then Group of the Wopmay orogen are not clearly delineated in the Great Lakes region, but late-stage acid volcanics and associated intrusives are followed by southerly transported quartzites in several areas south of Lake Superior (Dott and Dalziel, 1972). The general absence of these late-stage deposits in the eastern part of the Southern Province may be due to later uplift and erosion during the Grenville orogeny. The margins of Lake Superior must also have undergone considerable uplift during the opening of the Keweenawan rift system.

LaRue and Sloss (1980) argued against an aulacogen model because the early Proterozoic succession of the Great Lakes region records a general deepening of the depositional basin. Such a deepening upward succession is, however, present in other aulacogens (Hoffman and others, 1974). The record of the final phase of emergent-to-shallow water deposition is largely missing from the Great Lakes region.

4. Folding in the Lake Superior-Lake Huron region is mainly parallel to the basin axis as in other aulacogens. Although there is no evidence of large-scale nappe tectonics such as described by Hoffman and others (1977) from the Athapuscow aulacogen, early deformation and uplift of the axial zone may be indicated by the McGregor phase of the Penokean orogeny.

5. There are few plutonic igneous rocks in the Penokean fold belt. This is typical of aulacogens and contrasts with the abundant plutonic processes common in subduction-related orogens.

6. The Great Lakes aulacogen is situated on a major structural discontinuity (Sims and others, 1980), the Great Lakes Tectonic Zone. It has been suggested (Warner, 1978) that this structure may be part of a major Precambrian fault zone, the Colorado Lineament, extending across the North American continent. Warner (ibid.) proposed that it was a transcurrent fault system like the McDonald Fault in the Athapuscow aulacogen (Hoffman and others, 1974; Hoffman, 1980).

7. The Huronian outcrop belt is characterised by a negative gravity anomaly (O'Hara and Hinze, 1980) as are other aulacogens.

A comparison of the Athapuscow and Great Lakes aulacogens is shown in Figure 6. To facilitate visual comparison, the Great Slave Lake region is shown as a mirror image.

OTHER EARLY PROTEROZOIC AULACOGENS IN THE CANADIAN SHIELD

In addition to the well exposed and studied Athapuscow aulacogen in the Great Slave Lake region, Hoffman (1980) and Campbell (1981) proposed the existence of a second aulacogen about 800 km to the north in the vicinity of Coronation Gulf and the southwest end of Victoria Island (Fig. 7). Recently, Chandler and Schwarz (1980) proposed the existence of an aulacogen in the Richmond Gulf area on the east side of Hudson Bay. This feature was interpreted as opening to the west, into the region presently occupied by the Belcher fold belt. The early clastic fill of the aulacogen was transported eastward, towards the ancient continental interior. A small graben-like feature on the west side of the Labrador Trough at Cambrien Lake (Fig. 7) was interpreted by Burke (1980) as an aulacogen.

If, as suggested here, the northern Great Lakes region was the site of an early Proterozoic aulacogen, the available radiometric dates suggest that the structure was initiated earlier than the others described above. According to Hoffman (1980), the Athapuscow aulacogen came into being less than 2.1 Ga ago as indicated by dates obtained from rift-parallel diabase dykes and alkaline complexes. These diabase dykes have similar ages to the Nipissing diabase which was intruded during, or slightly later than, the McGregor phase of the Penokean orogeny in the Lake Huron area. Diabase dykes of about 2.1 Ga old are very widespread; they are known from the western part of the Canadian Shield to Labrador, Greenland, and Scandinavia (Payne and others, 1964). These intrusions may reflect a tensional episode of continental dimensions that culminated in continental fragmentation about 2.0 Ga ago. Earlier, more localised(?) attempts at continental breakup are recorded in the Huronian succession and possibly in the Wilson Island Group in the Great Slave Lake area (Hoffman, 1980). Reactivation of older structures plays an important role in localisation of many depositional basins. For example, the depositional trough in which the Wilson Island Group was deposited became the Athapuscow aulacogen. The early Proterozoic Great Lakes aulacogen developed on the site of the Archean Great Lakes Tectonic Zone (Sims and others, 1980) and was subsequently the site of the Keweenawan rift.

The present sites of several great lakes in the Canadian Shield are zones occupied by fault-related Proterozoic depositional basins. In northwestern Canada, the East Arm of Great Slave Lake mirrors closely the location of the Athapuscow aulacogen. Great Bear Lake may lie close to a westerly extension of the Taktu aulacogen of Campbell (1981). Lake Superior and the northern part of Lake Huron overlie the proposed site of the Great Lakes aulacogen and the younger Keweenawan trough. A major lake also overlies the aulacogen described by Chandler and Schwarz (1980) in the Richmond Gulf area. The longevity of some of these features points out the importance of reactivation of ancient structures in the localisation of fault-related depositional basins.

CONCLUSIONS

The upper part of the Huronian Supergroup is considered to be correlative with the lower part (Chocolay Group) of the Marquette Range Supergroup. These rocks were deposited from a dominantly eastward-flowing pa-

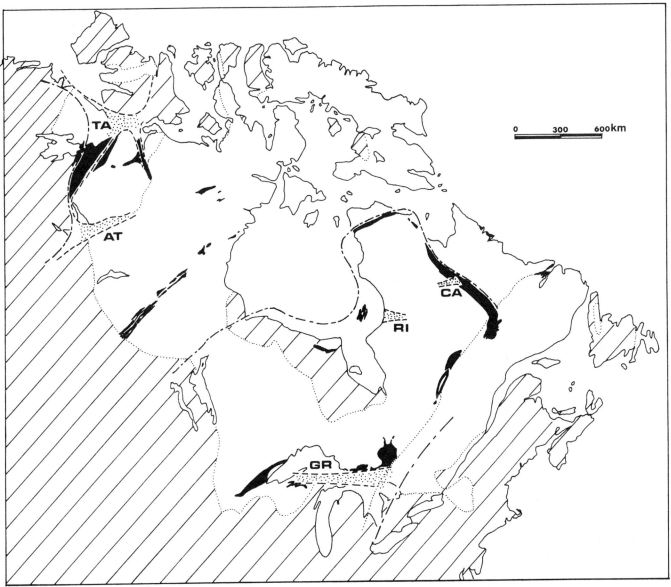

Figure 7. Sketch map of the northern part of North America to show locations of some areas of early Proterozoic supracrustal rocks (black ornament). Areas that have been interpreted as aulacogens are indicated by dotted ornament and are designated as follows: TA—Taktu aulacogen (Campbell, 1981), AT—Athapuscow aulacogen (Hoffman, 1980), CA—Cambrien Lake (Burke, 1980), RI—Richmond Gulf area (Chandler and Schwarz, 1980), and GR—the Great Lakes aulacogen (this paper).

leocurrent regime. Following some folding and uplift (McGregor phase), the iron-formations of the Superior basin were deposited. The succeeding depositional episode consists of a westerly-prograding flysch apron, the proximal facies of which is represented by the Whitewater Group of the Sudbury Basin. All of these rocks were then affected by the Michigan-Sudbury phase of the Penokean orogeny. The time of intrusion of the Sudbury Irruptive relative to these fold phases is not clearly defined.

The tectono-sedimentary history of the northern Great Lakes region in the early Proterozoic suggests that the area was the site of an aulacogen that opened towards the east. Several such aulacogens may have existed in the Canadian Shield between about 2.5 and 1.7 Ga ago. The longevity of these fault-bounded troughs is illustrated by subsequent development of younger depositional basins in the same areas.

ACKNOWLEDGMENTS

Financial assistance was provided by N.S.E.R.C. I wish also to express my thanks to the numerous graduate students at the University of Western Ontario who have contributed to the understanding of the Huronian and to the officers of the Ontario Geological Survey and the Geological Survey of Canada. Special thanks are due to W. R. Church who read an earlier version of this manuscript and made substantial contributions to the final paper. Responsibility for the interpretations must, however, rest solely with the author.

REFERENCES CITED

Aldrich, L. T., Davis, G. L., and James, H. L., 1965, Ages of minerals from metamorphic and igneous rocks near Iron Mountain, Michigan: Journal of Petrology, v. 6, p. 445–472.

Anderson, J. L., Van Schmus, W. R., and Medaris, L. G., 1975, Proterozoic granitic plutonism in the Lake Superior region and its tectonic implications [abs.]: EOS (American Geophysical Union Transactions), v. 56, p. 603.

Banks, P. O., and Van Schmus, W. R., 1971, Chronology of Precambrian rocks of Iron and Dickinson Counties, Michigan [abs.]: 17th Annual Institute on Lake Superior Geology, p. 9–10.

——1972, Chronology of Precambrian rocks of Iron and Dickinson Counties, Michigan, Part II [abs.]: 18th Annual Institute on Lake Superior Geology, p. 23.

Brocoum, S. J., and Dalziel, I.W.D., 1974, The Sudbury Basin, the Southern Province, the Grenville Front and the Penokean orogeny: Geological Society of America Bulletin, v. 85, p. 1571–1580.

Burke, K., 1980, Intracratonic rifts and aulacogens, in Continental Tectonics: National Research Council, Studies in Geophysics, p. 42–49.

Cambray, F. W., 1977, The geology of the Marquette District, a field guide: Michigan Basin Geological Society, 62 p.

——1978, Plate tectonics as a model for the environment of deposition and deformation of the early Proterozoic (Precambrian X) of northern Michigan: Geological Society of America Abstracts with Programs, v. 10, p. 376.

Campbell, F.H.A., 1981, Stratigraphy and tectono-depositional relationships of the Proterozoic rocks of the Hadley Bay area, northern Victoria Island, District of Keewatin: Geological Survey of Canada, Paper 81–1A, p. 15–22.

Cannon, W. F., 1973, The Penokean orogeny in northern Michigan, in Young, G. M., ed., Huronian stratigraphy and sedimentation: Geological Association of Canada Special Paper 12, p. 251–271.

Cantin, R., and Walker, R. G., 1972, Was the Sudbury Basin circular during deposition of the Chelmsford Formation?, in Guy-Bray, J. V., ed., New developments in Sudbury Geology: Geological Association of Canada Special Paper 10, p. 93–101.

Card, K. D., 1976, Geology of the McGregor Bay-Bay of Islands area, Districts of Sudbury and Manitoulin, Ontario: Ontario Division of Mines, Geological Report 138, 63 p.

——1978a, Geology of the Sudbury-Manitoulin area, Districts of Sudbury and Manitoulin: Ontario Geological Survey, 238 p.

——1978b, Metamorphism of the middle Precambrian (Aphebian) rocks of the eastern Southern Province, in Fraser, J. A., and Heywood, W. W., eds., Metamorphism in the Canadian Shield: Geological Survey of Canada, Paper 78–10, p. 269–282.

——1979, The Creighton pluton, Ontario: an unusual example of a forcefully emplaced intrusion: discussion: Canadian Journal of Earth Sciences, v. 16, p. 2181–2182.

Card, K. D., and Hutchinson, R. W., 1972, The Sudbury structure: its regional geological setting, in Guy-Bray, J. V., ed., New developments in Sudbury geology: Geological Association of Canada Special Paper 10, p. 67–78.

Chandler, F. W., 1973, Clastic dykes at Whitefish Falls, Ontario and the base of the Huronian Gowganda Formation, in Young, G. M., ed., Huronian stratigraphy and sedimentation: Geological Association of Canada Special Paper 12, p. 199–209.

Chandler, F. W., and Schwarz, E. J., 1980, Tectonics of the Richmond Gulf area, northern Quebec—a hypothesis: Geological Survey of Canada, Paper 80–1C, p. 59–68.

Chandler, F. W., Young, G. M., and Wood, J., 1969, Diaspore in early Proterozoic quartzites (Lorrain Formation) of Ontario: Canadian Journal of Earth Sciences, v. 6, p. 337–340.

Church, W. R., 1966, The status of the Penokean orogeny in Ontario [abs.]: Ninth conference on Great Lakes research, Chicago, p. 25.

——1967, The occurrence of kyanite, andalusite and kaolinite in lower Proterozoic (Huronian) rocks of Ontario [abs.]: Geological Association of Canada, Abstracts of Papers, p. 14–15.

——1968, The Penokean and Hudsonian orogenies in the Great Lakes region and the age of the Grenville Front [abs.]: 14th Annual Institute on Lake Superior Geology, p. 16–18.

——1971, Nature and evolution of Proterozoic and Phanerozoic orogenic belts [abs.]: Geological Association of Canada, p. 14–15.

Church, W. R., and Young, G. M., 1970, Discussion of the progress report of the Federal-Provincial Committee on Huronian stratigraphy: Canadian Journal of Earth Sciences, v. 7, p. 912–918.

——1972, Precambrian Geology of the Southern Canadian Shield with emphasis on the lower Proterozoic (Huronian) of the north shore of Lake Huron: International Geological Congress 24th, Montreal, Guidebook to field excursion A36–C36, 65 p.

Dietz, R. S., and Holden, J. C., 1966, Miogeoclines (miogeosynclines) in space and time: Journal of Geology, v. 74, p. 566–583.

Dott, R. H. Jr., and Dalziel, I.W.D., 1972, Age and correlation of the Precambrian Baraboo quartzite of Wisconsin: Journal of Geology, v. 80, p. 552–568.

Dutch, S. I., 1979, The Creighton pluton, Ontario: an unusual example of a forcefully emplaced intrusion: Canadian Journal of Earth Sciences, v. 16, p. 333–349.

Fairbairn, H. W., Hurley, P. M., and Pinson, W. H., 1960, Mineral and rock ages at Sudbury-Blind River, Ontario: Proceedings of the Geological Association of Canada, v. 12, p. 41–46.

Fairbairn, H. W., Hurley, P. M., Card, K. D., and Knight, C. J., 1969, Correlation of radiometric ages of Nipissing diabase and Huronian metasediments with Proterozoic orogenic events in Ontario: Canadian Journal of Earth Sciences, v. 6, p. 489–497.

Fleet, M. E., 1979, Tectonic origin for Sudbury, Ontario, shatter cones: Geological Society of America Bulletin, Part I, v. 90, p. 1177–1182.

——1980, Tectonic origin for Sudbury, Ontario, shatter cones: Reply: Geological Society of America Bulletin, Part I, v. 91, p. 755–756.

Frarey, M. J., Loveridge, W. D., and Sullivan, R. W., 1982, A U-Pb zircon age for the Creighton granite, Ontario: Geological Survey of Canada, Paper 82–1C, p. 129–132.

Frarey, M. J., and Roscoe, S. M., 1970, The Huronian Supergroup, north of Lake Huron, in Baer, A. E., ed., Symposium on basins and geosynclines of the Canadian Shield: Geological Survey of Canada, Paper 70–40, p. 143–157.

Gair, J. E., 1975, Bedrock geology and ore deposits of the Palmer Quadrangle, Marquette County, Michigan, with a section on the Empire mine by T. M. Han: U.S. Geological Survey Professional Paper 769, 159 p.

Gibbins, W. A., Adams, C. J., and McNutt, R. H., 1972, Rb-Sr isotopic studies of the Murray granite, in Guy-Bray, J. V., ed., New developments in Sudbury geology: Geological Association of Canada Special Paper 10, p. 61–66.

Gibbins, W. A., and McNutt, R. H., 1972, Rubidium-strontium studies on the Sudbury Irruptive: Geological Society of America Abstracts with Programs, v. 4, p. 517.

——1975, The age of the Sudbury Nickel Irruptive and the Murray granite: Canadian Journal of Earth Sciences, v. 12, p. 1970–1989.

Goldich, S. S., Nier, A. E., Baadsgaard, H., Hoffman, J. H., and Krueger, H. W., 1961, The Precambrian geology and geochronology of Minnesota: Minnesota Geological Survey Bulletin 41, 193 p.

Guy-Bray, J. V., 1972, ed., New developments in Sudbury geology: Geological Association of Canada Special Paper 10, 124 p.

Hanson, G. N., and Malhotra, R., 1971, K-Ar ages of mafic dikes and evidence for low-grade metamorphism in northeastern Minnesota: Geological Society of America Bulletin, v. 82, p. 1107–1114.

Hoffman, P. F., 1980, Wopmay orogen: a Wilson cycle of early Proterozoic age in the northwest of the Canadian Shield, *in* Strangway, D. W., ed., The continental crust and its mineral deposits: Geological Association of Canada Special Paper 20, p. 523–549.

Hoffman, P. F., Dewey, J. F., and Burke, K., 1974, Aulacogens and their genetic relation to geosynclines, with a Proterozoic example from Great Slave Lake, Canada, *in* Dott, R. H. Jr., and Shaver, R. H., eds., Modern and ancient geosynclinal sedimentation: Society of Economic Paleontologists and Mineralogists, Special Paper 19, p. 30–55.

Hofmann, H. J., Pearson, D.A.B., and Wilson, B. H., 1980, Stromatolites and fenestral fabric in early Proterozoic Huronian Supergroup, Ontario: Canadian Journal of Earth Sciences, v. 17, p. 1351–1357.

Innes, D. G., and Colvine, A. C., 1979, Metallogenetic development of the eastern part of the Southern province of Ontario. Summary of Field work, 1979: Ontario Geological Survey, Miscellaneous Paper 90, p. 184–189.

James, H. L., 1955, Zones of regional metamorphism in the Precambrian of northern Michigan: Geological Society of America Bulletin, v. 66, p. 1455–1488.

James, H. L., and Clark, L. D., Lamey, C. A., and Pettijohn, F. J., 1961, Geology of central Dickinson County, Michigan: U.S. Geological Survey Professional Paper 310, 176 p.

Keighin, C. W., Morey, G. B., and Goldich, S. S., 1972, East-central Minnesota, *in* Sims, P. K., and Morey, G. B., eds., Geology of Minnesota: a centennial volume: Minnesota Geological Survey, p. 240–255.

Knight, C. J., 1967, Rubidium-strontium isochron ages of volcanic rocks of the North Shore of Lake Huron [Ph.D. thesis]: University of Toronto.

Krogh, T. E., and Davis, G. L., 1974, The age of the Sudbury Nickel Irruptive: Annual Report of the Geophysical Laboratory, Carnegie Institute, Washington, Yearbook, v. 73, p. 567–569.

LaRue, D. K., and Sloss, L. L., 1980, Early Proterozoic sedimentary basins of the Lake Superior region: summary: Geological Society of America Bulletin, Part I, v. 91, p. 450–452.

Long, D.G.F., 1977, Resedimented conglomerate of Huronian (lower Aphebian) age, from the north shore of Lake Huron, Ontario, Canada: Canadian Journal of Earth Sciences, v. 14, p. 2495–2509.

——1978, Depositional environment of a thick Proterozoic sandstone: the (Huronian) Mississagi Formation, Ontario, Canada: Canadian Journal of Earth Sciences, v. 15, p. 190–206.

Lumbers, S. B., 1975, Geology of the Burwash area, Districts of Nipissing, Parry Sound and Sudbury: Ontario Division of Mines, Geological Report 115, 160 p.

Maas, R. S., Medaris, L. G. Jr., and Van Schmus, W. R., 1980, Penokean deformation in central Wisconsin, *in* Morey, G. B., and Hanson, G. N., eds., Selected studies of Archean gneisses and lower Proterozoic rocks, southern Canadian Shield: Geological Society of America, Special Paper 182, p. 147–157.

Morey, G. B., 1972, Gunflint Range, *in* Sims, P. K., and Morey, G. B., eds., Geology of Minnesota: a centennial volume: Minnesota Geological Survey, p. 218–225.

——1973, Stratigraphic framework of middle Precambrian rocks in Minnesota, *in* Young, G. M., ed., Huronian stratigraphy and sedimentation, Minnesota, *in* Young, G. M., ed., Huronian stratigraphy and sedimentation, Geological Association of Canada Special Paper 12, p. 211–249.

Morey, G. B., and Sims, P. K., 1976, Boundary between two Precambrian W terranes in Minnesota and its geological significance: Geological Society of America Bulletin, v. 87, p. 141–152.

Morris, W. A., 1980, Tectonic and metamorphic history of the Sudbury Norite: the evidence from Paleomagnetism: Economic Geology, v. 75, p. 260–277.

O'Hara, N., and Hinze, W. J., 1980, Regional basement geology of Lake Huron: Geological Society of America Bulletin, Part I, v. 91, p. 348–358.

Parviainen, A.E.U., 1973, The sedimentation of the Huronian Ramsay Lake and Bruce Formations, north shore of Lake Huron [Ph.D. thesis]: University of Western Ontario, 426 p.

Payne, A. V., Baadsgaard, H., Burwash, R. A., Cumming, G. L., Evans, C. R., and Folinsbee, R. E., 1964, A line of evidence supporting continental drift, *in* Smith, C. H., and Sorzenfrei, T., eds.: The upper mantle symposium, New Delhi, 1964, Copenhagen, p. 83–90.

Peredery, W. V., 1972, Chemistry of fluidal glasses and melt bodies in the Onaping Formation, *in* Guy-Bray, J. V., ed., New developments in Sudbury geology: Geological Association of Canada Special Paper 10, p. 49–59.

Pettijohn, F. J., 1943, Basal Huronian conglomerates of the Menominee and Calumet districts, Michigan: Journal of Geology, v. 51, p. 387–397.

——1970, The Canadian Shield—a status report, 1970, *in* Baer, A. J., ed., Symposium on basins and geosynclines of the Canadian Shield: Geological Survey of Canada, Paper 70–40, p. 248–257.

Puffett, W. P., 1969, The Reany Creek Formation, Marquette County, Michigan: U.S. Geological Survey Bulletin 1274-F, 25 p.

Robertson, J. A., 1973, A review of recently acquired geological data, Blind River-Elliot Lake area, *in* Young, G. M., ed., Huronian stratigraphy and sedimentation: Geological Association of Canada Special Paper 12, p. 169–198.

Roscoe, S. M., 1969, Huronian rocks and uraniferous conglomerates in the Canadian Shield: Geological Survey of Canada Paper 68–40, 205 p.

——1973, The Huronian Supergroup, a paleoaphebian succession showing evidence of atmospheric evolution: Geological Association of Canada Special Paper 12, p. 31–47.

Rousell, D. H., 1972, The Chelmsford Formation of the Sudbury Basin—a Precambrian turbidite, *in* Guy-Bray, J. V., ed., New developments in Sudbury geology: Geological Association of Canada Special Paper 10, p. 79–91.

Sawiuk, M. J., 1977, The Thessalon metavolcanics; a consideration of the geology, petrography, and geochemistry [B.Sc. thesis]: University of Western Ontario, 97 p.

Schwerdtner, W. M., and Lumbers, S., 1980, Major diapiric structures in the Superior and Grenville Provinces of the Canadian Shield, *in* Strangway, D. W., ed., The continental crust and its mineral deposits: Geological Association of Canada Special Paper 20, p. 149–180.

Sims, P. K., 1976, Precambrian tectonics and mineral deposits, Lake Superior region: Economic Geology, v. 71, p. 1092–1127.

Sims, P. K., Card, K. D., Morey, G. B., and Peterman, Z. E., 1980, The Great Lakes tectonic zone—a major crustal structure in central North America: Geological Society of America Bulletin, Part I, v. 91, p. 690–698.

Smith, E. I., 1978, Precambrian rhyolites and granites in south-central Wisconsin: field relations and geochemistry: Geological Society of America Bulletin, v. 89, p. 875–890.

Stevenson, J. S., 1972, The Onaping ash-flow sheet, Sudbury, Ontario, *in* Guy-Bray, J. V., ed., New developments in Sudbury geology: Geological Association of Canada Special Paper 10, p. 41–48.

Van Schmus, W. R., 1965, The geochronology of the Blind River-Bruce Mines area, Ontario, Canada: Journal of Geology, v. 73, p. 755–780.

Van Schmus, W. R., 1976, Early and middle Proterozoic history of the Great Lakes area, North America: Philosophical Transactions of the Royal Society of London, v. A280, p. 605–628.

——1980, Chronology of igneous rocks associated with the Penokean orogeny in Wisconsin, *in* Morey, G. B., and Hanson, G. N., eds., Selected studies of Archean gneisses and lower Proterozoic rocks, southern Canadian Shield: Geological Society of America, Special Paper 182, p. 159–168.

Van Schmus, W. R., and Anderson, J. L., 1977, Gneiss and migmatite of Archean age in the Precambrian basement of central Wisconsin: Geology, v. 5, p. 45–48.

Van Schmus, W. R., Card, K. D., and Harrower, K. L., 1975, Geology and ages of buried basement rocks, Manitoulin Island, Ontario: Canadian Journal of Earth Sciences, v. 12, p. 1175–1189.

Warner, L. A., 1978, The Colorado lineament: a middle Precambrian wrench fault system: Geological Society of America Bulletin, v. 89, p. 161–171.

Wetherill, G. W., Davis, G. L., and Tilton, G. R., 1960, Age measurements on minerals from the Cutler batholith, Cutler, Ontario: Journal of Geophysical Research, v. 65, p. 2461–2466.

Wood, J., 1973, Stratigraphy and depositional environments of upper Huronian rocks of the Rawhide Lake-Flack Lake area, Ontario, *in* Young, G. M., ed., Huronian stratigraphy and sedimentation: Geological Association of Canada Special Paper 12, p. 73–95.

Young, G. M., 1966, Huronian stratigraphy of the McGregor Bay area, Ontario; relevance to the paleogeography of the Lake Superior region: Canadian Journal of Earth Sciences, v. 3, p. 203–210.

——1970, An extensive early Proterozoic glaciation in North America?: Palaeogeography, Palaeoclimatology, Palaeocology, v. 7, p. 85–101.

——1971, Stratigraphic and sedimentological framework of the Southern province of the Canadian Shield [abs.]: Geological Association of Canada Annual Meeting, Sudbury, p. 75–76.

——1972, Downward intrusive breccias in the Huronian Espanola Formation, Ontario, Canada: Canadian Journal of Earth Sciences, v. 9, p. 756–762.

——1973a, Origin of carbonate-rich early Proterozoic Espanola Formation, Ontario, Canada: Geological Society of America Bulletin, v. 84, p. 135–160.

——1973b, Tillites and aluminous quartzites as possible time markers for middle Precambrian (Aphebian) rocks of North America, *in* Young, G. M., ed., Huronian stratigraphy and sedimentation: Geological Association of Canada Special Paper 12, p. 97–127.

——1981, The early Proterozoic Gowganda Formation, *in* Hambrey, M. J., and Harland, W. B., eds., Pre-Pleistocene glacial record on Earth. Cambridge University Press, p. 807–812.

Young, G. M., Long, D.G.F., and McLennan, S. M., 1977, Deltaic deposits in the upper Pecors, Espanola and Gowganda Formations (Huronian) [abs.]: Abstracts and Proceedings 23rd Annual Meeting of Lake Superior Institute, p. 46.

MANUSCRIPT ACCEPTED BY THE SOCIETY MARCH 4, 1983

Geological Society of America
Memoir 160
1983

Early Proterozoic tectonics of the Lake Superior region: Tectonostratigraphic terranes near the purported collision zone

D. K. Larue
Department of Geology
Stanford University
Stanford, CA 94305

ABSTRACT

Two assemblages of early Proterozoic rocks are present in the Lake Superior region: a northern miogeoclinal assemblage of stratified rocks overlying Archean basement; and a southern magmatic arc terrane composed of granitoid and volcanic rocks. The structural contact between the miogeoclinal assemblage and the magmatic terrane in northeastern Wisconsin and upper Michigan is the Florence-Niagara fault. Miogeoclinal assemblage rocks directly north of the Florence-Niagara fault in the study area are subdivided into two fault-bound, tectonostratigraphic terranes: the Crystal Falls terrane (or terrane 1); and the Florence-Niagara terrane (or terrane 3). These structural blocks are segregated from the miogeocline proper (terrane 2) to the north because of differences in sedimentary facies, and/or structure. I interpret the Penokean Orogeny (1.8-1.9 b.y.) as a product of the collision between the magmatic terrane and the miogeocline: terranes 1 and 3 probably were emplaced during or after this collision but may have been discrete from the miogeocline prior to collision. It is probable that the early Proterozoic evolution of the Lake Superior region is much like that of Phanerozoic orogens, but gross similarities are obfuscated by the presence of, and the deformation associated with emplacement of, tectonostratigraphic terranes, which occur both in the Phanerozoic and the Precambrian.

INTRODUCTION

Belts of highly deformed rock are relatively common features in all early Proterozoic terranes, yet their origins remain unclear. Many recent studies have concentrated on comparing and contrasting these so-called mobile belts to the better understood Phanerozoic orogens. For example, Kroner (1977, 1981) argues that the early Proterozoic may have been characterized at least in part by intracontinental orogenies owing, perhaps, to A-type subduction (subduction of continental crust). Hoffman (1973, 1980) argues convincingly for Phanerozoic-style tectonic processes operating in the early Proterozoic of the Wopmay Orogeny.

One of the classic belts of deformed early Proterozoic rocks is the circum-Superior, rimming the Archean Superior province. Baragar and Scoates (1981) summarized the evolution of the belt and noted that the Lake Superior region (Fig. 1) is the only segment of the belt that exhibits

possible evidence of arc-type magmatism during the early Proterozoic. They concluded that this area may have been characterized by ocean-floor subduction and generation of granitic magmas.

The following will concentrate on the evolution of the Lake Superior segment of the circum-Superior belt during the early Proterozoic. The first part of the paper will review the geology of the Lake Superior region and the several hypotheses proposed to explain its complex history. This first part of the paper stresses that a collision best explains the present-day juxtaposition of a magmatic arc terrane with an approximately coeval to older miogeocline. In the second part of the paper, the contact relations between the magmatic terrane and the miogeocline are analyzed, and it is concluded that two terranes consisting of miogeoclinal lithologies are present between the miogeocline and the

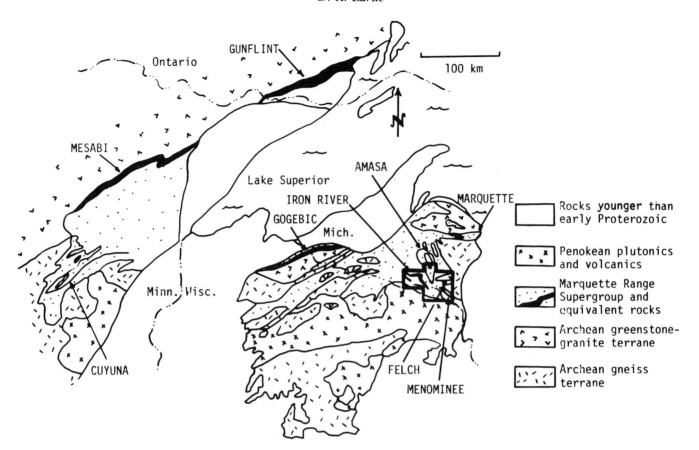

Figure 1. Location map of Lake Superior region. Boxed area shows location of Figure 2 map.

magmatic terrane: these two terranes are named the Crystal Falls terrane and the Florence-Niagara terrane. The Crystal Falls terrane, miogeocline, and Florence-Niagara terranes are numbered terranes 1, 2, and 3, respectively, to differentiate similar lithologies separated by faults. Each of the three terranes is characterized by subtly different stratigraphic successions, constituent facies, and structures. Archean basement is exposed only in the miogeocline, terrane 2. (Note added in proof: "Archean" rocks shown in the magmatic terrane of Fig. 1 have been shown to be of early Proterozoic age; P. K. Sims, personal communication, 1982).

In the conclusion of the paper, it is suggested that the early Proterozoic evolution of the Lake Superior region is indeed much like that of Phanerozoic orogens, but the gross similarities are obfuscated somewhat by the presence of, and the deformation associated with emplacement of, tectonostratigraphic terranes, which occur both in the Phanerozoic and in the Precambrian.

Regional Geology

Before describing results of recent work in the vicinity of the proposed suture zone, a brief review of relevant Lake Superior geology is necessary.

Figure 1 is a geologic map of the Lake Superior region and shows that there are three important units that require discussion: 1) Precambrian basement older than early Proterozoic (mostly Archean); 2) early Proterozoic Marquette Range Supergroup (1.85-2.0 b.y.; Van Schmus, 1980); 3) magmatic (arc?) terrane (1.8-1.85 m.y.; Van Schmus, 1980). Deformation of these rocks occurred during the Penokean Orogeny, about 1.8-1.85 b.y. ago (Van Schmus, 1980). The study area considered in the second part of this paper is boxed in Figure 1.

Where Marquette Range Supergroup strata overlie Archean crystalline basement, the rocks are termed miogeoclinal. Fault-bounded packets of Marquette Range Supergroup rock present between the magmatic terrane and areas definitely underlain by Archean crystalline rock in the study area (Fig. 1) have been segregated from the miogeocline for reasons discussed later and are termed the Crystal Falls terrane and Florence-Niagara terrane (Fig. 2). The miogeocline and the two terranes are referred to as the miogeoclinal assemblage.

Older Precambrian Basement. Older Precambrian basement can be divided into two types: Huronian strata (2.2-2.3 b.y.) and Archean (>2.4 b.y.) crystalline rocks (Van Schmus, 1976). Huronian strata are relatively rare in the Lake Superior region (Van Schmus, 1976) and are not

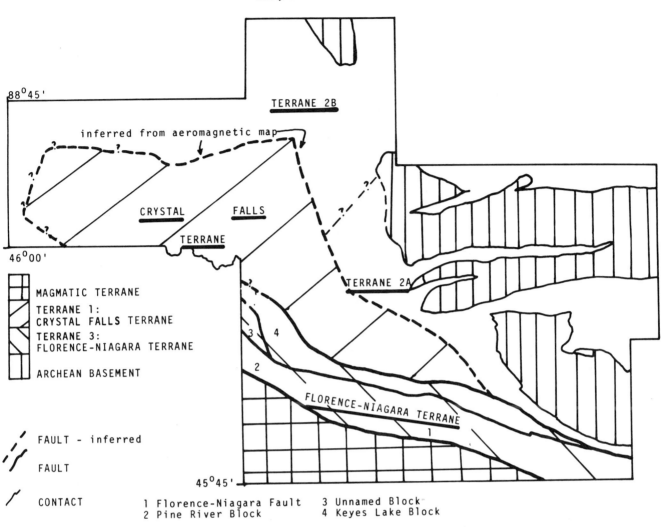

Figure 2A. Terrane map of study area showing: Crystal Falls terrane (terrane 1); terranes 2A and 2B; Florence-Niagara terrane (terrane 3), with relation to magmatic terrane and Archean basement. See text. Note that faults defining Crystal Falls terrane are *inferred* based on interpretation of the aeromagnetic map of Zietz and Kirby (1971). Numbers refer to Florence-Niagara fault and to fault blocks in Florence-Niagara terrane. Dickinson area mentioned in text—Felch and Calumet trough areas. Crystal Falls is located in the NE corner of Crystal Falls terrane.

discussed here. Morey and Sims (1976) divided the Lake Superior region Archean rocks into a younger (2.7-2.5 b.y.), northern, granite-greenstone terrane and an older (>3.0 b.y.), southern, gneissic terrane (Fig. 1).

Marquette Range Supergroup. The Chocolay Group (probable equivalent of the Mille Lacs Group in Minnesota; Morey, 1978a), consisting of basal conglomerate overlain successively by quartzite, dolomite, and slate, is the lowermost unit of the Marquette Range Supergroup. Larue (1979, 1981a) and Larue and Sloss (1980) presented evidence that the Chocolay Group sediments accumulated in rift basins and on platforms between rift basins.

Menominee Group (lower part of Animikie Group in Minnesota) strata rest unconformably on Chocolay Group strata, and consist of quartzite or muddy quartzite overlain by banded iron formation. In the Marquette trough, the thick Siamo Slate occupies a position between the quartzite and iron formation. Because the Siamo Slate is a turbidite unit in part (Larue, 1979, 1981b) and because it is restricted to the Marquette trough, it was concluded that the Marquette trough was an actively subsiding basin during deposition of Menominee Group sediments.

The Baraga Group overlies the Menominee Group unconformably in Michigan and Wisconsin, whereas its equivalents in Minnesota and Ontario (the upper part of the Animikie Group) rest conformably on underlying banded iron formation. The Baraga Group and equivalents are dominantly deep-water turbiditic-pelitic units. In the

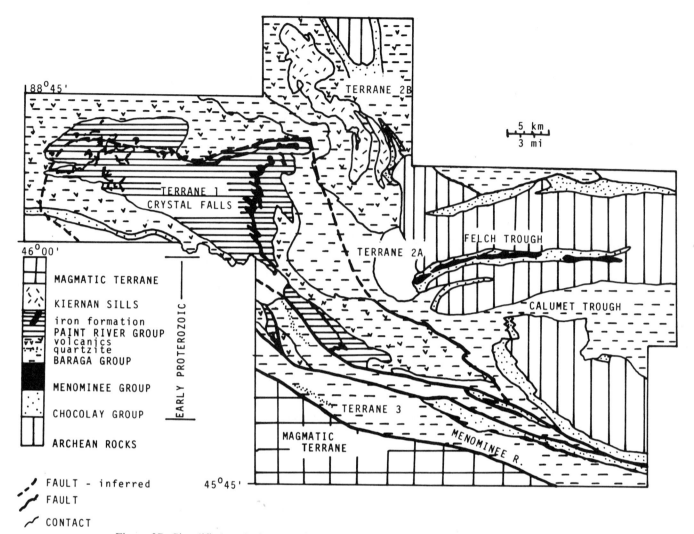

Figure 2B. Simplified geologic map of study area (after Gair and Weir, 1956; Bayley, 1959; James and others, 1966; Weir, 1967; James and others, 1968; Dutton, 1971).

Iron River-Crystal Falls area, thick mafic volcanic flows are abundant in the Baraga Group. Metarhyolitic tuffs are rare in the Crystal Falls-Dickinson County area.

Alwin (1979) and Larue and Sloss (1980) proposed that an elongate sedimentary basin, termed the Animikie basin, contains Animikie Group and some of the equivalent (Menominee and Baraga Group) strata. The Animikie basin was bound approximately by the Gunflint and Mesabi Ranges on its northwest side (Morey and Ojakangas, 1970) and the Gogebic Range on the southeast side (Alwin, 1979; Larue and Sloss, 1980) (see Fig. 1). Another deepwater basin existed southeast of the Animikie basin during deposition of other Baraga Group strata, and its boundaries are not known.

The Paint River Group is composed of slate, graywacke, and iron formation lithologies and overlies great thicknesses of Baraga Group pillow basalts (see Fig. 2B). Unlike the aforementioned groups, the Paint River Group

occurs exclusively in the Crystal Falls and Florence areas and nowhere else in the Lake Superior region (what will later be called the Crystal Falls terrane and Florence-Niagara terrane). James and others (1968) argued that the Paint River Group is stratigraphically the youngest early Proterozoic succession in the Lake Superior region. Cambray (1978) suggested that the Paint River Group strata and underlying pillow basalts are the equivalent of the Menominee Group. The pillow basalts below the Paint River Group were mapped as part of the Baraga Group by James and others (1968) and Dutton (1971). The argument of Cambray (1978) is *not* preferred here because putting the pillow basalt unit at the base of the Menominee Group, as he suggests, means that these rocks would be stratigraphically equivalent to blanket quartzites of the Menominee Group, which seems unlikely.

Early Proterozoic mafic intrusive rocks are commonly found intruding Marquette Range Supergroup strata. They have not been dated, but locally intrude all major groups, and are deformed by Penokean events.

Larue and Sloss (1980) concluded that there were two major periods of basinal subsidence during Marquette Range Supergroup sedimentation: the first occurred during Chocolay Group sedimentation; the second occurred during deposition of the Menominee-Baraga-Paint River Group sediments. They considered the first subsidence epoch to be related to rifting and formation of oceanic crust probably to the south. The second period of basin subsidence is not well understood but probably related to a second rifting event or onset of strike-slip faulting. The second period of basin subsidence established region-wide, deep-water sedimentation.

Penokean Orogeny. Following formation of this diverse assemblage of sedimentary and volcanic rocks, the rocks were deformed and regionally metamorphosed during the "Penokean Orogeny" (1.8-1.85 b.y.; see Van Schmus, 1980). Although orientations of fold axes show large variance (Cannon, 1973), the intensity of deformation decreases in a northerly direction. Mapping of metamorphic isograds (James, 1955) in the Lake Superior region has revealed the presence of annular nodes ranging from greenschist grade at the periphery of each node to as high as sillimanite at the center. Mean metamorphic grade, like intensity of deformation, also appears to decrease toward the north.

Sims (1976; personal communication, 1982) has stressed the importance of gneiss doming in both the miogeoclinal terrane and the magmatic terrane during the Penokean Orogeny. Gneiss domes in the miogeocline are cored with Archean basement whereas gneiss domes in the magmatic terrane are cored by early Proterozoic gneisses.

Proposed Interpretation of Tectonics. Based on James' (1954) analysis of the early Proterozoic history of the Lake Superior region in terms of geosynclinal evolution, several recent theories have been proposed which attempt to clarify and update James' work.

1) Morey and Sims (Morey, 1973; Morey and Sims, 1976; Sims, 1976; Morey, 1978b; Sims, 1980; Sims and others, 1980) suggest that the entire early Proterozoic history of the region can be explained in terms of intracratonic deposition in a large interior basin and subsequent intracratonic deformation.

2) Van Schmus (1976) interpreted the Chocolay and Menominee Groups (and equivalents) as possibly indicating passive-margin deposition, whereas the Baraga Group strata were perhaps deposited in a foreland basin. He tentatively modeled the Penokean Orogeny as a product of a consuming continental margin, with ocean floor (to the south) subducted toward the north, under the foreland basin.

3) Cambray (1977, 1978) has suggested a complex tectonic history of the early Proterozoic of the Lake Superior region. He has suggested that the Chocolay Group strata were deposited in a cratonic setting and that the strata of the Menominee Group show features which are inferred to indicate initiation of rifting. In this model, the Penokean Orogeny was caused by the collision of a continent to the north (bearing the Marquette Range Supergroup strata) with an arc to the south (containing Archean rocks in central Wisconsin as described by Van Schmus and Anderson, 1977). The suture extends W to WNW across northern Wisconsin and is delimited by Penokean granitic plutons. The Florence-Niagara fault (Fig. 2) is thought to be the actual suture in the Cambray model for the present study area (the fault has other names along its strike).

EVIDENCE OF A SUTURE AND COLLISION OROGEN

This section contains a partial list of why the Florence-Niagara fault (Fig. 2) should or should not be considered a suture and why the Penokean Orogeny should or should not be considered a collision orogen. Reasons to doubt the suture/collision interpretation are: 1) absence of ophiolites, melange or blueschist (paired metamorphic belts); and 2) lack of observed thrust faults in the Lake Superior region (foreland, accretionary-prism, or collision-produced). Major suture belts and collision zones throughout the world generally show the above characteristics. However, many sutures in the western United States lack some of these lithologic features due to retrograde metamorphism and disruption or removal during late stage strike-slip faulting or uplift. The lack of observed thrust faults in the Lake Superior region is problematic. However, cryptic thrust faults have been proposed by Cannon (1973) and Klasner (1978). Holst (in preparation) has documented the existence of an early recumbent fold set in Minnesota in Baraga Group equivalent strata, which he suggests may have been formed during northward thrusting. The lack of observed thrust faults in the Lake Superior region may be because: 1) they were not formed during the Penokean Orogeny; 2) deformed rocks presently observed were deformed *below* a major decollement surface; 3) thrust faults are present but are cryptic—lack of minor thrusts on outcrop scale is countered by presence of major thrusts which take up most displacement.

Evidence that the Florence-Niagara fault may be a suture includes: 1) the dramatic change in chemistry, metamorphic grade, and structural style across the fault. Cudzilo (1978) showed that the fault separates igneous rocks of presumed continental tholeiitic nature (miogeocline) from calc-alkaline (magmatic terrane); 2) the fundamental change in stratigraphy and lithotypes across the boundary—the thinning or disappearance of unequivocal Marquette Range Supergroup strata to the south; 3) regional patterns of Marquette Range Supergroup sedimentation are best interpreted (Larue and Sloss, 1980) in terms of a rifted continental margin (thus supporting a miogeoclinal setting).

SEDIMENTARY AND STRUCTURAL TERRANES IN SOUTHCENTRAL UPPER MICHIGAN AND ADJACENT WISCONSIN

Cambray (1978) and Larue and Sloss (1980) both agree that the contact between the magmatic terrane and the miogeoclinal assemblage, the Florence-Niagara fault, is the most likely location for the early Proterozoic suture (Fig. 2A, B). Preliminary results presented herein concern the geology of the Marquette Range Supergroup immediately adjacent to and north of the Florence-Niagara fault. Studies of the structures in the magmatic rocks south of the proposed suture are not yet completed (R. L. Sedlock, in preparation).

Based on differences in stratigraphy and structural fabrics, three terranes are recognized in the study area (Fig. 2) north of the proposed suture. Terrane 1, the Crystal Falls terrane (CFt), encompasses the Iron River-Crystal Falls district; terrane 2, east of terrane 1, is underlain by Archean basement and possesses stratigraphic and structural features similar to many other areas in the southern Lake Superior region; terrane 3, referred to as the Florence-Niagara terrane (FNt), parallels the trace of the proposed suture (Fig. 2A, B), The CFt and the FNt both contain stratigraphic and structural features that are different from those observed in other early Proterozoic rocks in the Lake Superior region.

Crystal Falls Terrane

Boundaries and Distinguishing Features. The CFt is roughly equidimensional in shape and is inferred to be fault-bound. The CFt is in fault contact on the south with the FNt (Fig. 2A). An inferred fault to the east separates the CFt from terrane 2A, and an inferred fault to the north separates the CFt from terrane 2B (the latter two fault boundaries are inferred based on truncation of aeromagnetic trends shown on the map by Zietz and Kirby, 1971). There is insufficient outcrop to define the exact boundaries of the southwest and west sides of terrane 1; however, faults are inferred based on preliminary study of aeromagnetic anomalies.

Stratigraphy. The oldest early Proterozoic sediments exposed in the CFt are silicified dolomites and quartzose terrigenous rocks of the Chocolay Group, which are poorly exposed in the southern part of this terrane (Fig. 2A, B). No unequivocal Menominee Group rocks are present in the CFt. Instead, Baraga Group extrusive volcanics (Badwater Greenstone) may unconformably (?) overlie Chocolay Group sediments (James and others, 1968) or may be in

fault contact. Turbiditic sandstones and slates are present structurally beneath the Badwater Greenstone elsewhere in terrane 1, indicating that the contact with this carbonate is probably a fault. Presumably deep-water turbiditic sandstones and pelites of the Paint River Group unconformably (?) overlie Baraga Group strata (James and others, 1968). A deep-water (below storm-wave base) origin is proposed because of the fine-grained, thin-bedded nature of these rocks; their significant lateral extent without evidence of interfingering with shallow-water deposits; and their significant formation thicknesses within sequences that show no evidence of shallowing.

Structure. The structure of these rocks has not yet been studied in detail (W. L. Ueng, in preparation). James and others (1968) presented evidence that complex deformation has occurred. Fold axes in the area are mostly NW-trending and have been openly refolded by a younger event; locally W- and NE-trending major folds are observed (James and others, 1968).

Subterrane 2A

Terrane 2 is divided into subterranes 2A and 2B (Fig. 2), which have different geologic features but are not apparently separated by any major structural discontinuity. Terrane 2 is the only terrane discussed here that is demonstrably floored by Archean basement.

Boundaries and Distinguishing Features. Subterrane 2A is inferred to be fault-bound on the west with the CFt and fault-bound on the south with the FNt, is overlain to the east by Phanerozoic cover, and is apparently continuous to the northwest (subterrane 2B) and to the north. Subterrane 2A is unique with regard to the other terranes discussed here because it lacks the thick sequence of Baraga Group volcanics of the CFt and subterrane 2B, and significant mappable units in the Baraga Group that show the mature (quartzose and carbonate) lithologies of the FNt.

Stratigraphy. The oldest Proterozoic rocks of subterrane 2A are those of the Chocolay Group, which include locally thick accumulations of quartzite and dolomite. Muddy quartzites and iron formation of the Menominee Group unconformably overlie strata of the Chocolay Group and are relatively thin. Baraga Group slates and dirty sandstones, locally with volcaniclastic interbeds (James and others, 1968), cap the sequence.

Larue and Sloss (1980) and Larue (1981a) have suggested that the Felch and Menominee structural troughs were sedimentary basins during Chocolay Group sedimentation. This interpretation is based on the presence of trough-parallel paleocurrent vectors from cross-beds in quartzite in the areas within the structural troughs. The Felch basin was bound on either side by platforms, now expressed by exposed Archean basement. Archean basement is present only on the northeast side of the Menominee trough. The southwest margin of the Menominee trough is structurally complicated and not well understood. In fact, this side of the basin may have been open, such that the Menominee basin may have had only one well-defined margin.

During Menominee Group sedimentation, the Felch and Menominee basins received little sediment. Apparently, the basins became inactive during this period, and it is not known whether these basins were reactivated during Baraga Group sedimentation. The complex structural and metamorphic overprinting of these strata precludes detailed sedimentologic analysis. Baraga Group sediments represent products of regional deep-water sedimentation.

Structure. In the Felch trough, major folds have E-W axes that are parallel to the trend of the trough (James and others, 1961).

South of the Felch trough in the Calumet trough, highly deformed and metamorphosed Baraga Group slates and schists are foliated: poles to foliations are distributed in a great circle about a shallowly plunging, ENE-trending axis. This fabric is indicative of coaxial refolding about a shallow, ENE-trending axis (Fig. 3A).

Some of the Calumet trough bedding is homoclinal, NW-striking, and quasivertical.

West of the Calumet trough, in the Michigamme Dam area, NW-trending folds with shallow plunges and steeply dipping axial planes are present. The folds are, locally, refolded by open folds with near-vertical axes (Fig. 3B, C).

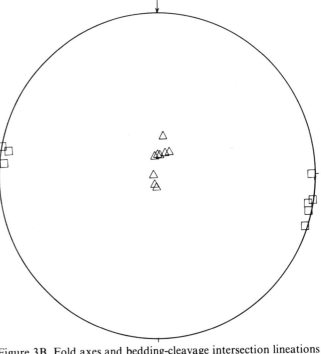

Figure 3B. Fold axes and bedding-cleavage intersection lineations (triangles) and poles to axial planes of minor folds and foliations (squares) for Michigamme Slate; south and southwest of Peavy Pond (on terrane 1/2 boundary; which is not well-defined here). Location: hollow symbols. R31W T42N, Sec 32SE; R31W T41N, Sec. 4. Filled symbols, R31W T42N, Sec 31, 32 W.

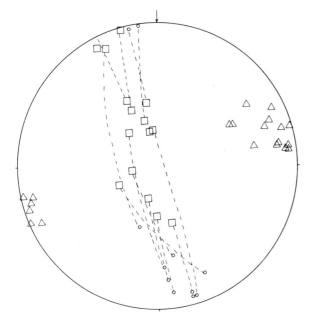

Figure 3A. Fold axes of minor folds and bedding-cleavage intersections (triangles) and poles to foliations and axial planes of minor folds (squares) from Michigamme Slate in Calumet trough area. Poles to foliations are tied to bedding poles to clarify orientations. Location: R28W T41N, Sec 19, 20, 21; R29W T41N, Sec 24.

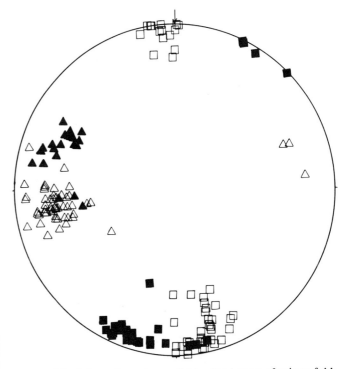

Figure 3C. Orientation of cross-folds. Fold axes of minor folds are shown as triangles; poles to axial planes of minor folds are shown as squares. Same location as 3B.

Thus, because of this refolding, it is possible that these folds are of the same generation as those in the Calumet trough but have been reoriented by younger folding. The difference in the mean orientation of their axes is attributed to later open refolding about steep axes. No evidence of coaxial refolding is observed in this area, however, suggested coaxial refolding may have been significant only locally.

Subterrane 2B

Subterrane 2B is different from subterrane 2A because the former contains much greater volumes of early Proterozoic mafic igneous rocks and was subject to a more complicated deformation history than the latter.

Stratigraphy. The basement of subterrane 2B is Archean gneiss which forms the core of a large dome, the southern tip of which is shown in Figure 2A. Marquette Range Supergroup strata mantle the dome. Chocolay Group dolomites comprise the lowermost part of the sedimentary drape, and are overlain by Baraga Group volcanics and sediment. Of particular interest is the Kiernan sills intrusive complex, which is of greatest thickness to the southwest of the domes (Fig. 2B) but is probably equivalent to scattered, smaller mafic intrusions that occupy a similar stratigraphic position around the perimeter of the dome (Cannon, 1978). Where it is thickest, the Kiernan complex resembles an ophiolite sequence and consists of: 1) a cumulate-layered ultramafic rock base (100-300 m); 2) a cumulate and isotopic gabbro medial section (1-3 km); and 3) a pillow basalt cap (>1 km) overlain by ribbon cherts and turbiditic ash beds (>100 m), and iron formation (the Mansfield member; ~200 m) (descriptions from Gair and Weir, 1956; Bayley, 1959; Weir, 1967). Obviously, the Kiernan complex does not represent a fragment of ancient ocean crust if it conformably overlies Archean crystalline basement; however, the Kiernan complex conceivably could have been tectonically juxtaposed over the Chocolay Group strata. The Kiernan complex is being subject to an intensive investigation to test this hypothesis (Fox and others, in preparation); preliminary results favor an autochthonous origin as the floor of a volcanic basin within continental crust.

Structure. The structure of subterrane 2B has not been studied in detail (W. L. Ueng, in preparation). However, the gneiss dome shown in Figure 2B is considered to be younger than the major folding of terrane 1, 2A, and 3 rocks. The dome is elongate approximately N-S, and both the bedding and the cleavage in the sedimentary mantle wrap around the oval (Fig. 4). This foliation, which probably formed perpendicular to the direction of regional shortening (Wood, 1974), either is: 1) an earlier foliation that has been refolded around the oval; or 2) was produced during formation of the oval. Because the oval is elongated N-S and because youngest folds with N-S axial planes are

observed in areas not obviously related to the oval (discussed in terrane 2A, 3 structural section), it is tentatively concluded that the dome was produced during a period of E-W shortening. The processes which are inferred to have led to the formation of the dome remain unclear; however, the E-W shortening that was in some way related to dome formation also affected rocks in other terranes (terranes 2A, 3) and thus dates the oval formation as post major folding.

Florence-Niagara Terrane

The FNt consists of at least three prominent fault slices containing Marquette Range Supergroup strata, forming steeply to moderately S-SSE-dipping and -facing homoclines (Figs. 2A, B). The three fault slices are the Pine River block (PRb), the unnamed block (ub), and the Keyes Lake block (KLb) (Fig. 2A). These rocks were most recently studied by Bayley and others (1966), who studied the eastern area, and by Dutton (1970), who mapped the western area. Lithologies and structure of each block are discussed below.

Pine River Block. The PRb contains Chocolay (Randville Dolomite only), Menominee, and Baraga Group (Michigamme Slate only) strata. The lithology and structures of these strata are discussed sequentially.

The Randville Dolomite consists of the following lithic types:

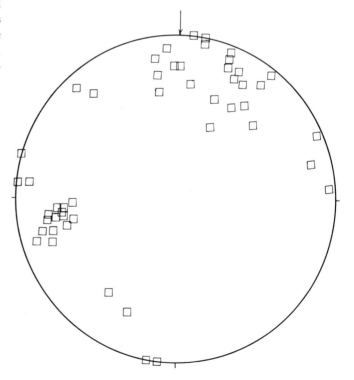

Figure 4. Poles to foliations from Kiernan area (southern tip of gneiss dome). Data from Gair and Weir (1956).

1) massive, cryptalgal laminated and stromatolitic dolomicrospar;

2) sandy dolomicrospar (dominantly in the lower part of the section);

3) rare quartzitic and arkosic sandstone beds 10-100 cm thick;

4) silicified oolitic dolomicrospar;

5) rare siliciclastic mud interbeds 1-3 cm thick.

Lithologies 1-3 are locally silicified. Bayley and others (1966) and Larue (1979, 1981a) described sedimentary structures found in these rocks and concluded deposition occurred in subtidal to intertidal environments on a carbonate platform, with local exposures of Archean basement supplying siliciclastic detritus.

For the purpose of this study, the Menominee Group can be considered to be composed of Felch Formation, a muddy quartzite unit, overlain by the Vulcan Iron Formation. The Vulcan Iron Formation has been shown to consist of several members (Bayley and others, 1966) which are not considered here. The Felch Formation is nowhere seen in contact with the underlying Randville Dolomite, and pebbles of Randville Dolomite are *not* present in the Felch in this area. Therefore, the contact relations between the Felch and the Randville are not understood, although most workers *prefer* to consider it an unconformity based on regional relations known elsewhere in the Lake Superior region which may or may not apply here.

The Felch Formation consists of interlayered sandstones and mudstones from 1-200 cm thick. The conformably overlying Vulcan Iron Formation is composed of banded chert-hematite beds, and interstratified pelitic horizons (see also Bayley and others, 1966, for more details on lithology).

The Michigamme Slate in the eastern PRb consists of the following lithotypes: 1) thinly-bedded to massive pelites with and without interbedded feldspathic sandstones and siltstones (commonly graded); 2) massive to medium-bedded quartzites interbedded with pelites or dolomite (see below). Quartzite beds are rarely graded; 3) plane-laminated to massive dolomites with and without interbedded pelitic horizons. The Michigamme Slate in the eastern PRb is thought to be composed primarily of lithotype (1); the resistant nature of lithotypes (2) and (3) makes ridges composed of these lithologies relatively common. These sediments were probably deposited in some type of submarine fan setting, as indicated by the interstratified graded sandstones and mudstone. The dolomites may indicate some sort of slope deposit, although this is speculation based on little data. However, no evidence of shallow-water deposition was seen in these rocks.

The Michigamme Slate of the western part of the PRb consists of two basic lithotypes: 1) pebbly quartzitic sandstone and associated quartzites and conglomerates of shallow-water origin (Nilsen, 1965); and 2) pelitic rocks, thin-bedded sandstones, graphitic slate, and agglomeratic units of uncertain but probable deep-water origin. The first lithotype occupies a medial position between two belts of the second lithotype (Fig. 2B). The contact between the two lithotypes may be structural.

Unnamed Block (Ub). The unnamed block is unique with regard to the other blocks described in the FNt in that it contains Badwater Greenstone (Baraga Group) and Paint River Group strata (the Dunn Creek Slate and the overlying Riverton Iron Formation strata). The Badwater Greenstone contains pillow basalts and is structurally massive; it is not considered further herein. The Dunn Creek Slate consists of pelites and interstratified thin graded beds composed of graywacke. The Riverton Iron Formation consists largely of interbedded chert-hematite layers 1-10 cm thick.

Keyes Lake Block. The Keyes Lake Block (KLb) consists of Chocolay Group strata (Randville Dolomite only) overlain by Menominee Group and Baraga Group strata (Bayley and others, 1966).

The Randville Dolomite in the KLb consists of two lithotypes: 1) massive, laminated or stromatolite-bearing dolomicrospar, commonly silicified; and 2) conglomerate consisting of subrounded clasts of lithotype (1) above in a mud matrix of comminuted dolomicrospar and sandy terrigenous material. The second lithotype, mapped as the basal conglomerate of the Menominee Group (Bayley) and others, 1966) can be shown to be interstratified with lithotype (1) above (Larue, 1979), and is interpreted to be a channel-fill deposit.

The Menominee Group in the KLb consists of the Felch Formation overlain by the Vulcan Formation. The Felch Formation has been mapped in this area as consisting dominantly of silicified breccia, quartzite, and slate. The silicified breccia shows *no* evidence of being of detrital origin, and thus may be either a silcrete deposit on the sub-Menominee Group unconformity or a younger hydrothermal deposit. Thus, silicified breccia is considered herein *not* to be a member of the Felch Formation. This conclusion is further supported by the lack of silicified breccia in the Felch Formation elsewhere in other blocks in the FNt. Quartzite and slate lithologies in the Felch Formation are extremely rare in the KLb.

The lithologies in the Vulcan Iron Formation are similar to those described in the PRb.

The Baraga Group of the KLb consists of Michigamme Slate and Badwater Greenstone. The Michigamme Slate is poorly exposed in the eastern KLb, and only three outcrop areas have been mapped (Bayley and others, 1966). Two of those exposures are very poor, and are massive beds of sandstone and intercalated mudstone. The third exposure (T39N, R31E, sec. 24NE) was mapped by Bayley and others (1966) as muddy Chocolay Group quartzite (Sturgeon Quartzite) but was reinterpreted by Larue (1979) as Michigamme Slate. The outcrop consists of thin beds of sandstone with interstratified mud; several of the sand-stones show features interpreted here to indicate turbidity-current origin (e.g., ripple laminations grading into mud).

The western part of the KLb contains Michigamme Slate and Badwater Greenstone. The Badwater Greenstone is massive, is poorly exposed, and contains local interpillow sediments; it is not discussed further herein. The Michigamme Formation is composed of two principal litholo-gies: 1) massive pelite with thin sandy interbeds; and 2) massive quartzite with local pebbly horizons. Lithology type (1) is possibly indicative of deep-water sedimentation, although this conclusion is speculative owing to lack of diagnostic sedimentary structures. The massive quartzite unit, showing a maximum mapped outcrop width of 0.5 km, consists of five sublithotypes: a) structureless quartzite; b) plane-laminated quartzite; c) tabular cross-bedded quartzite with set thickness 10-50 cm; d) structureless peb-bly quartzite; and e) massive quartzite with interbedded pelitic horizons, 1-3 cm thick. Nilsen (1965) concluded that the quartzite unit was deposited in a shallow marine setting, and I concur with him. The contact between the major lithotypes is probably structural.

Significance of Lithology. The FNt differs significantly in lithology with terranes 1 and 2 only with respect to the Michigamme Slate. Differences are: 1) presence of signifi-cant volumes of dolomite in the Michigamme Slate in the eastern PRb; and 2) presence of shallow-water quartzite lenses in both the PRb and KLb (the Goodrich Quartzite in the Marquette area, the basal unit of the Baraga Group, is somewhat similar to these quartzites; however, the FNt quartzites occur within the sequence). Because of the fault-bound nature of the FNt and its different stratigraphy, the unit was segregated from the miogeocline and termed a separate terrane.

Structure. Two episodes of deformation are recog-nized in the FNt. The first deformation was a major event, and styles of deformation associated with this deformation show great diversity. Strain in the rocks resulting from this deformation is highly variable, ranging from no apparent grain deformation, to great finite strain in which grains and pebbles show extreme elongation (Fig. 5A). In two or per-haps three areas in the complicated belt, fold axes of minor folds and bedding-cleavage intersections are distributed in a great circle about a constant axial plane (Fig. 5B). Minor folds show moderate thickening in hinge regions and open to small apical angles (0-120°). Such a distribution of fold axes may be explained either as: 1) a result of great strain following or accompanying a single deformation event (Escher and Watterson, 1974; Ramsay, 1979) or 2) a result of two phases of deformation, the first phase being cryptic and the second phase superposed on an already folded sur-face. Several lines of evidence lead me to suspect that the fabric reflects significant finite strain in the rocks: 1) stretched pebbles such as those present in terrane 3 are not

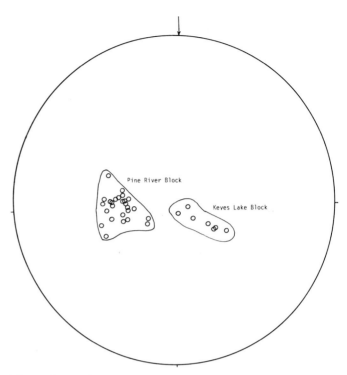

Figure 5A. Orientations of long axes of stretched pebbles in quartzite units, terrane 3.

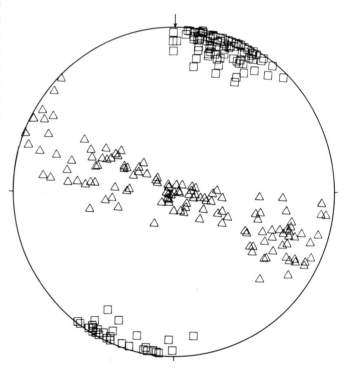

Figure 5B. Summary diagram, terrane 3 (Florence-Niagara ter-rane). Fold axes of minor folds and lineations shown as triangles; poles to axial planes of minor folds and cleavage shown as squares. Location of data: cleaved slates and quartzites in the Pine River and Keyes Lake block (see location, Fig. 2), and south of Hanbury Lake (R29W T39N, Sec 15, 16).

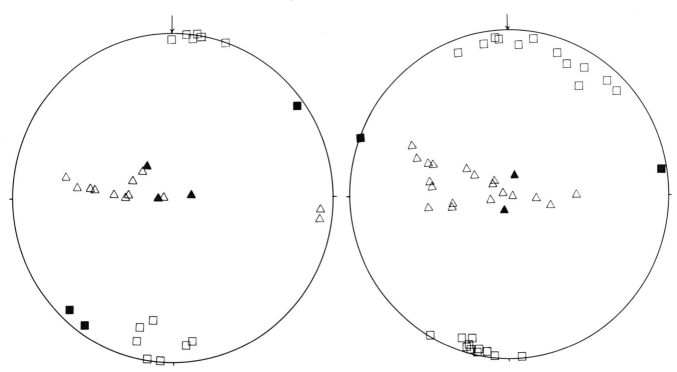

Figure 5C. Fold axes of minor folds and bedding-cleavage intersection lineations (triangles) and poles of axial planes of minor folds and cleavage (squares) in Michigamme Slate, of older (hollow symbols) and younger (filled symbols) deformation ages. Location: R20E T39N, Sec 4; R30W T39N, Sec. 11.

Figure 5D. Fold axes of minor folds and bedding-cleavage intersection lineations (triangles) and poles of axial planes of minor folds and cleavage (squares) in Riverton Iron Formation, of older (hollow symbols) and younger (filled symbols) deformation ages. Location: R17E T40N, Sec 35, and R17E T39N, Sec 2 (Bessie Babbet Lake area).

present in other domains, suggesting this region has undergone greater strain; 2) thickening in the hinges of minor folds with attenuation of limbs is significant in terrane 3 folds, but rare in folds from other domains; and 3) apical angles in folds from the complicated belt tend to be less than those from the other domains, although this point has not yet been quantified. In other areas of the FNt, only partial girdles or clusters of folds axes are observed (Fig. 5C). Such simple fabrics may indicate that their folds axes were more closely aligned with the axis of the strain ellipsoid, and thus did not undergo rotation during strain. Strain data from these rocks and fold style observations are still inconclusive with respect to differentiating between these arguments.

The nature of the second deformation in terrane 3 is uncertain. In Figure 5D, the second deformation is oriented nearly N-S and openly folds the F_1 structures. Minor kink folds with N-S axial planes and associated open folding are present throughout the FNt, and I suggest this is the same pulse of deformation as the cross-folds shown in Figure 5D. Interestingly, these younger folds have axial traces that are subparallel to the axial trace of the dome in subterrane 2B, tentatively suggesting perhaps that the deformation event responsible for the dome may have produced folds which

are not clearly related to the structure of the oval in nearby rocks.

SUMMARY OF THE SEDIMENTARY-TECTONIC EVOLUTION OF TERRANES 1, 2, AND 3

Chocolay Group sediments were deposited in fault-bound basins and on platforms overlying Archean basement in terrane 2. Less is known about these sediments in terranes 1 or 3, although a similar origin is likely. Menominee Group strata are thin in terranes 2 and 3, and absent in the CFt: no evidence of local basin subsidence associated with sedimentation is observed in any of the three terranes during Menominee Group sedimentation. Baraga Group strata in terrane 1 include several kilometers of extrusive volcanics, whereas mineralogically mature to immature sediments and lesser amounts of volcanics are present in the FNt. Subterrane 2A contains both mature and immature sediments with minor volcanics. Terranes 1 and 2 show no evidence of shallow-water sedimentation during Baraga Group deposition (other than at the basal unconformity of the Baraga Group in terrane 2) and consist of probable deep-water turbidites and pelites. Terrane 3 contains some quartzites of apparent shallow-water origin interleaved with probable deep-water turbidites and pelites. The Paint River Group is found only in terranes 1 and 3, and probably represents deep-water deposition (James and others, 1968).

In summary, it is clear that the three terranes are characterized by similar stratigraphic sequences and tectonostratigraphic histories but also possess subtle differences that necessitate their separation into discrete domains.

BEARING ON NATURE OF EARLY PROTEROZOIC TECTONICS

At this point, it is necessary to summarize features of the three lithotectonic terranes. First, terrane 2 is fault-bounded with respect to terranes 1 and 3 but apparently *not* with respect to other rocks in the Lake Superior region. Thus, terrane 2 is autochthonous; there has probably been no movement of terrane 2 with respect to the rest of the interior of the ancient continent. That is, events occurring in terrane 2 are in some way related to regional events that occurred elsewhere on the autochthonous continental margin. However, because terranes 1 and 3 are stratigraphically different from terrane 2 and are *fault-bound* (note that fault-boundaries are only inferred for the CFt), one cannot *assume* that the terranes are autochthonous with respect to the continental interior (see, for example, Coney and others, 1980). However, because terranes 1 and 3 contain miogeoclinal lithologies, they can be considered parautochthonous, or not necessarily far-travelled. Any attempt to reconstruct the sedimentary and tectonic evolution of the area must take the above points into consideration.

There are three major ways in which the sedimentary and tectonic evolution of the three terranes and the adjacent magmatic terrane can be interpreted. First, the entire succession could represent a product of intracratonic or pericratonic deformation, and the faults bounding the terranes do not have large displacements along them. Using this interpretation, a complex history of sedimentation, characterized by two periods of basin subsidence in a cratonic or pericratonic environment, was followed by deformation. Such deformation could have been produced by continental margin activity occurring further to the south (back-arc deformation, for example), or by some sort of uniquely Precambrian intracratonic deformation mechanism.

Note that using this argument, the magmatic terrane is related genetically to the Marquette Range Supergroup, albeit probably not time-correlative. This interpretation would suggest that the area evolved from a cratonal or pericratonal setting to one dominated by arc-type volcanism. Such a change could have been caused by the initiation of subduction beneath the Lake Superior region around 1.9 b.y. ago (Van Schmus, 1976). The Florence-

Niagara fault in this case would be a back-arc strike-slip (?) fault. An example of intracratonic or pericratonic deformation occurred in the western United States during the early Tertiary Laramide orogeny and is thought to have been produced by low-angle subduction (Dickinson, 1978). Note, however, that in this case sedimentation is not dominated by deep-water deposition as in the Lake Superior region.

The second possible hypothesis for the sedimentary and tectonic evolution of the area is that of Cambray (1978), as modified by Larue and Sloss (1980). In this scenario, a passive margin that has undergone two phases of basinal subsidence (a rift-related phase during Chocolay Group deposition and a poorly understood phase during Menominee-Baraga-Paint River Group sedimentation) collides with an arc system or active continental margin moving from south to north (present-day coordinates). This interpretation suggests that the Florence-Niagara fault is an interplate suture.

The third proposed model is new and a modification of hypothesis 2. This final idea emphasizes the fact that the southern Lake Superior region consist of both in-place units (terrane 2) and fault-bound terranes of unknown origin (magmatic terrane, terranes 1 and 3). The fault-bound terranes composed of miogeoclinal assemblage lithologies may have had two different origins. First, they may have become fault-bound *prior* to the orogeny, or purported collision. Second, following plate collision, a major phase of strike-slip faulting transported lithotectonic terranes of different stratigraphic character and locally re-oriented structures. In both respects, the Penokean terranes of the southern Lake Superior region may resemble the terranes recently described in the western United States, or the post-collision collage of fault-bounded slices in the Himalayas.

In summary, geologic evidence at this stage cannot be used to *conclusively* disprove any of the three hypotheses stated above. Instead, more data are needed on the chemistry and age of the various igneous rocks in the area, ages and types of metamorphism, and geometries and phases of deformation to understand the purported collision history of the belt. Furthermore, although it is generally accepted that the Marquette Range Supergroup represents a passive continental margin or intracratonic assemblage, causes of multiple basin subsidence are largely unknown. These and other problems in the area can be dealt with only by detailed field studies backed by laboratory analysis.

The following is a speculative reconstruction of sedimentation and tectonism that occurred during the early Proterozoic of the Lake Superior region (Fig. 6). In the discussion of sedimentation, it is significant that all three terranes can be described using the same stratigraphic nomenclature, with only local modification. However, tectonic setting during sedimentation is understood only in

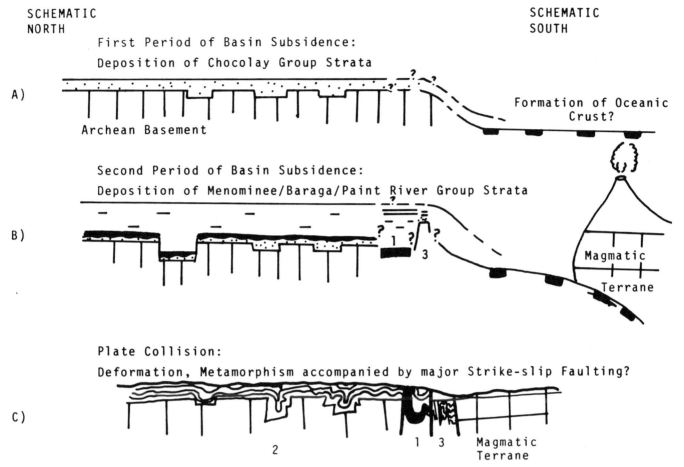

Figure 6. Speculative reconstruction showing evolution of study area from passive continental margin to collision zone. A) Deposition of Chocolay Group on continental crust dominantly in terrane 2, during rifting event. B) Deposition of Menominee-Baraga and Paint River Groups during second period of basin subsidence. Relations between terranes 1 and 3 (shown as numbers) and autochthonous continental margin now known. Basement of terrane 1 shown in black to signify unknown nature. C) Plate collision with accompanying deformation of continental margin. Major strike-slip faulting prior to and post regional metamorphism emplaces terranes 1 and 3 (see text for discussion).

the miogeocline (terrane 2), where regional stratigraphic relations can be established.

The Chocolay Group is considered to be a rift-related sequence, formed during N-S extension accompanying opening of an ocean basin somewhere in northern Wisconsin (Fig. 6A). The basal conglomeratic sediments of the Chocolay Group represent rift-related deposition; the uppermost sediments represent a miogeoclinal assemblage (blanket quartzites and dolomites). The region-wide unconformity between the Chocolay and Menominee Groups serves as evidence for two discrete epochs of sedimentation. During deposition of the Menominee Group strata, the passive continental margin became moderately activated, accompanied by local basin subsidence and volcanism (Larue, 1981b). This period of basinal subsidence and local uplift continued into Baraga Group sedimentation, at which time the entire margin foundered and deep-water

sedimentation became widespread throughout the Lake Superior region with the sole exception of the shallow-water quartzites in terrane 3 (these shallow-water quartzites are strange enough to make one hesitant about correlating these rocks with Baraga Group strata). This second phase of basin subsidence, certainly unrelated to the earlier Chocolay Group phase may represent: 1) a second phase of rifting (either further extending the continental margin or created during the formation of a back-arc basin; no back-arc basin-type sediments have been found to date, however); 2) transformation of the passive plate boundary into a conservative boundary (transcurrent). Either of these subsidence mechanisms is possible, but data are lacking to strongly support either case.

Following this complicated sedimentation history, the rocks were deformed in three major episodes. The first two events were coaxial and produced folds with axes trending shallowly WNW. The final phase was produced by E-W shortening and represents a product of gneiss doming, possibly caused by heating and subsequent mobilization of the Archean sialic crust during orogenesis. The first two deformation events are considered to be products of progressive deformation resulting from the collision of a southern arc or continent with a continental margin to the north bearing the Marquette Range Supergroup. The collision occurred slightly obliquely with respect to the continental margin, and strike-slip faulting ensued transporting, mixing, and reorienting fragments of both the arc system and the early Proterozoic sediments. The present-day collage of fault-bound terranes that are lithologically, stratigraphically, and structurally different to adjacent terranes is the product of this complicated sedimentary and tectonic history.

ACKNOWLEDGMENTS

This study was supported by NSF Grant EAR 80-08202 and NSF Grant EAR 81-08564. S. M. Karl, E. L. Miller, N. H. Sleep, W. L. Ueng, F. W. Cambray, and two anonymous reviewers read earlier versions of this manuscript.

Note added in final proof: Based on the map of Lake Superior region by Morey and others (1982), it appears likely that the Chocolay Group does *not* occur in terrane 1, but in the NW-extension of terrane 3, supporting further the contention that terrane 1 is discrete.

REFERENCES CITED

Alwin, Bevan, 1979, Sedimentology of the Tyler Formation [abs.]: Geological Society of America Abstracts with Programs, v. 11, no. 5, p. 225.

Baragar, W.R.A., and Scoates, R.F.J., 1981, The Circum-Superior belt: a Proterozoic Plate Margin?, *in* A. Kroner, ed., Precambrian Plate Tectonics, Elsevier, Amsterdam, p. 57–90.

Bayley, R. W., 1959, Geology of the Lake Mary Quadrangle, Iron County, Michigan: U.S. Geological Survey Bulletin, v. 1077, 112 p.

Bayley, R. W., Dutton, C. E., Lamey, C. A., and Treves, S. B., 1966, Geology of the Menominee iron-bearing district, Dickinson County, Michigan and Florence and Marinette Counties, Wisconsin: U.S. Geological Survey Professional Paper 513, 96 p.

Cambray, F. W., 1977, The geology of the Marquette District, A field guide: Michigan Basin Geological Society, 62 p.

—— 1978, Plate tectonics as a model for the environment of deposition and deformation of the early Proterozoic (Precambrian X) of northern Michigan: Geological Society of America Abstracts with Programs, v. 10, no. 7, p. 376.

Cannon, W. F., 1978, Geologic Map of the Iron River 1° x 2° Quadrangle Michigan and Wisconsin: U.S. Geological Survey Open File Map 78–342.

—— 1973, The Penokean Orogeny in northern Michigan, *in* Young, G. M., ed., Huronian stratigraphy and sedimentation: Geological Association of Canada Special Paper 12, p. 251–171.

Coney, P. J., Jones, D. L., and Monger, J. W., 1980, Cordilleran suspect terranes, Nature, v. 288, p. 329–333.

Cudzilo, T. F., 1978, Geochemistry of Early Proterozoic Igneous Rocks, Northeastern Wisconsin and Upper Michigan [Ph.D. thesis]: Univ. Kansas, Lawrence, 194 p., (unpublished).

Dutton, C. E., 1971, Geology of the Florence area, Wisconsin and Michigan: U.S. Geological Survey Professional Paper 633, p. 54.

Escher, A., and Watterson, J., 1974, Stretching fabrics, folds, and crustal shortening: Tectonophysics, v. 22, p. 223–231.

Gair, J. E., and Wier, K. L., 1956, Geology of the Kiernan Quadrangle, Iron County, Michigan: U.S. Geological Survey Bulletin, v. 1044, 88 p.

Goodwin, A. M., 1956, Facies relation in the Gunflint iron formation: Economic Geology, v. 51, p. 565–595.

Hoffman, P. F., 1973, Evolution of an Early Proterozoic continental margin: the Coronation geosyncline and associated aulacogens of the northwest Canadian Shield: Philos. Transactions Royal Society of London, Series A, 273: p. 547–561.

—— 1980, Wopmay Orogen: a Wilson Cycle of Early Proterozoic age in the northwest of the Canadian Shield: Geological Association Canada, Special Paper 20, p. 523–549.

James, H. L., 1954, Sedimentary facies of iron-formation: Economic Geology, v. 49, p. 235–294.

—— 1955, Zones of regional metamorphism in the Precambrian of northern Michigan: Geological Society of America Bulletin, v. 66, p. 1455–1487.

James, H. L., Clark, L. D., Lamey, C. L., and Pettijohn, F. J., 1961, Geology of central Dickinson County, Michigan: U.S. Geological Survey Professional Paper 310, 176 p.

James, H. L., and others, 1968, Geology and ore deposits of the Iron River-Crystal Falls district, Iron County, Michigan: U.S. Geological Survey Professional Paper 570, 134 p.

Klasner, J. S., 1978, Penokean deformation and associated metamorphism in the western Marquette Range, northern Michigan: Geological Society of America Bulletin, v. 89, p. 711–722.

Kroner, A., 1977, The Precambrian geotectonic evolution of Africa: plate accretion versus plate destruction, Precambrian Research, v. 4, p. 163–213.

—— 1981, Precambrian Plate tectonics, *in* A. Kroner, ed., Precambrian Plate Tectonics, Elsevier, Amsterdam, p. 57–90.

Larue, D. K., 1979, Sedimentary history prior to chemical iron sedimentation of the Precambrian X Chocolay and Menominee Groups (Lake Superior region) [Ph.D. thesis]: Northwestern University, 173 p.

—— 1981a, The Chocolay Group, Lake Superior region, U.S.A.: sedimentologic evidence for deposition in basinal and platform settings on an early Proterozoic craton: Geological Society of America Bulletin, v. 92, p. 417–435.

—— 1981b, The early Proterozoic Menominee Group siliciclastic sediments of the southern Lake Superior region: Evidence for sedimentation in platformal and basinal setting: Journal of Sedimentary Petrology, v. 51, p. 397–414.

Larue, D. K., and Sloss, L. L., 1980, Early Proterozoic sedimentary basins of the Lake Superior region: Geological Society of America Bulletin, Part I, v. 91, p. 450–452; Part II, v. 91, p. 1836–1874.

Morey, G. B., 1973, Stratigraphic framework of middle Precambrian rocks in Minnesota, *in* Young, G. M., ed., Huronian stratigraphy and sedimentation: Geological Association of Canada Special Publication 12, p. 211–249.

—— 1978a, Lower and middle Precambrian stratigraphic nomenclature for east-central Minnesota: Minnesota Geological Survey Report of Investigations 21, 52 p.

—— 1978b, Metamorphism in the Lake Superior region, U.S.A., and its relation to crustal evolution, *in* Metamorphism in the Canadian Shield: Geological Survey of Canada Paper 78–10, p. 283–319.

Morey, G. B., and Ojakangas, R. W., 1970, Sedimentology of the middle Precambrian Thomson Formation, east-central Minnesota: Minnesota Geological Survey Report of Investigations 13, 32 p.

Morey, G. B., and Sims, P. K., 1976, Boundary between two Precambrian W terranes in Minnesota and its geological significance: Geological Society of America Bulletin, v. 87, p. 141–152.

Nilsen, T. E., 1965, Sedimentology of middle Precambrian Animikian quartzites, Florence County, Wisconsin: Journal of Sedimentary Petrology, v. 35, p. 805–817.

Prinz, W. C., 1976, Correlative iron-formation and volcanic rocks of Precambrian age, Northern Michigan: 22nd Annual Institute on Lake Superior Geology, p. 47.

Ramsay, D. M., 1979, Analysis of rotation of folds during progressive deformation: Geological Society of America Bulletin, Part I, v. 90, p. 732–738.

Sims, P. K., 1976, Precambrian tectonics and mineral deposits, Lake Superior region: Economic Geology, v. 71, p. 1092–1127.

——1980, Boundary between Archean greenstone and gneiss terrane in northern Wisconsin and Michigan: Geological Society of America Special Paper 182, p. 113–124.

Sims, P. K., Card, K. D., Morey, G. B., and Peterman, Z. E., 1980, The Great Lakes tectonic zone—A major crustal structure in central North America: Geological Society of America Bulletin, Part I, v. 91, p. 690–698.

Van Schmus, W. R., 1976, Early and middle Proterozoic history of the Great Lakes area, North America: Royal Society of London Philosophical Transactions, Series A, v. 280, p. 605–628.

——1980, Chronology of igneous rocks associated with the Penokean Orogeny in Wisconsin: Geological Society of America, Special Paper 182, p. 159–168.

Van Schmus, W. R., Thurman, E. M., and Peterman, Z. E., 1975b, Geology and Rb-Sr chronology of middle Precambrian rocks in eastern and central Wisconsin: Geological Society of America Bulletin, v. 86, p. 1255–1265.

Van Schmus, W. R., and Anderson, J. L., 1977, Gneiss and migmatite of Archean age in the Precambrian basement of central Wisconsin: Geology, v. 5, p. 45–48.

Weir, D. S., 1967, Geology of the Kelso Junction Quadrangle, Iron County, Michigan: U.S. Geological Survey Bulletin, v. 1226, 47 p.

Wood, D. S., 1974, Current views of the development of slaty cleavage: Annual Review Earth Planetary Sciences, v. 2, p. 369–401.

Zietz, I., and Kirby, J. R., 1971, Aeromagnetic map of the western part of the northern peninsula of Michigan and part of northern Wisconsin: U.S. Geological Survey Map GP-750.

MANUSCRIPT ACCEPTED BY THE SOCIETY, MARCH 4, 1983

Printed in U.S.A.

Geological Society of America
Memoir 160
1983

Tidal deposits in the early Proterozoic basin of the Lake Superior region—The Palms and the Pokegama Formations: Evidence for subtidal-shelf deposition of Superior-type banded iron-formation

Richard W. Ojakangas
Department of Geology
University of Minnesota
Duluth, Minnesota 55812

ABSTRACT

The Palms Formation in Wisconsin and Michigan and the correlative Poke-gama Quartzite in Minnesota are interpreted as tidal deposits formed along the margins of the early Proterozoic Animikie basin.

The well-exposed Palms Formation, which extends for 85 km along the Gogebic range, is 150 m thick and can be divided into three units on the basis of rock types and bedding styles: (1) a thin lower unit of thin-bedded argillaceous rocks that unconformably overlies a low-relief surface of Archean granite and greenstone and older Proterozoic sedimentary units; (2) a thick middle unit of thin alternating beds of argillite, siltstone, and sandstone that vary considerably in texture and composition. Bedding types include parallel, wavy, cross, and flaser lamination, and a variety of sedimentary structures are present; (3) an upper unit of thicker beds of parallel and cross-bedded sandstone.

A total of 199 cross-bedding measurements and 52 measurements of other paleocurrent indicators from the Palms Formation yields a crude bimodal-bipolar distribution, with a broad primary mode to the west and a weaker mode to the east. The sandstones are mineralogically mature; most of the framework grains are rounded quartz grains. The sandstones of the middle unit have an abundant sericitic matrix, whereas those of the upper unit are texturally more mature.

The Pokegama Quartzite is exposed at only a few places along the 130 km long Mesabi range near the northwestern margin of the Animikie basin. However, the entire formation can be viewed in two drill cores in which it is 50 m and 26 m thick. Sedimentary sequences and the mineralogical attributes are similar to the Palms. The paleocurrent plot (N = 38) is crudely bimodal-bipolar with primary modes to the north and south.

By utilization of Walther's Law of Succession of Facies and comparisons with modern environments, it is postulable that both formations were deposited under transgressive tidal conditions. In this model, the lower (shaly) facies was deposited in a low-energy domain of the upper (shoreward) tidal flat; the middle facies (shale-siltstone-sandstone) was deposited on a middle tidal flat under alternating low- and high-energy conditions; and the upper facies (sandstone) was deposited in a lower tidal flat or subtidal high-energy environment.

The Palms and Pokegama formations pass upward into the Ironwood and Biwabik Iron Formations, respectively. Again using Walther's Law, it can be postulated that the iron-formations were deposited on a shelf located seaward from the subtidal sandstone facies. The "cherty" (coarser-grained, thicker-bedded, iron oxide-chert) facies was deposited in shallower water than was the "slaty" (finer-grained, thinner-bedded, iron silicate-iron carbonate) facies.

The tidal-subtidal facies model developed here provides an independent approach in evaluating the environment of deposition of one kind of Superior-type banded iron-formation. The model is primarily based upon the siliciclastic lithologies associated with iron-formation rather than upon the iron-formation itself.

INTRODUCTION

In early Proterozoic time, an elongate east-west trending basin developed astride the Great Lakes tectonic zone which originated during the Archean as a major structural feature (Sims and others, 1980). Pre-Animikie rocks deposited in this basin include the Mille Lacs Group in Minnesota (Morey, 1978 and 1979) and the Chocolay Group in Michigan and Wisconsin (Cannon and Gair, 1970). Broad, but gentle uplift resulted in partial erosion of the Chocolay and Mille Lacs strata, and as the region subsided once again, rocks of the Animikie and Menominee Groups were deposited unconformably upon these and Archean basement rocks (Table 1). The basin in which this younger sequence of Lower Proterozoic rocks accumulated is herein called the Animikie basin.

The broad outline of the western part of the Animikie basin is approximately delineated by several iron ranges (Fig. 1). Excellent exposures of Animikie strata are present on the Gogebic range, good artificial exposures are found on the Mesabi range, a few exposures exist on the Gunflint range, but only subsurface data are available from the Cuyuna range (Fig. 1).

The easternmost exposed strata of the Animikie basin occur in the Marquette range, where Menominee Group strata also were deposited upon the eroded Chocolay Group (Table 1). Southeastward from the Gogebic range, the Animikie basin was broader and included the Crystal Falls-Iron Mountain district (Dickinson County area) where the major stratigraphic relationships are generally similar to but more complex than those on the Marquette range. It has been suggested that a continuation of the basin may have extended eastward into Labrador (Goldich and others, 1961; Gross, 1965, p. 120; Dott and Batten, 1971; Baragar and Scoates, 1981). Paleozoic rocks cover the region to the east of Marquette, and where lower Proterozoic rocks reappear 300 km to the east, they are Huronian in age, older than the Animikie but perhaps

TABLE 1.   CORRELATION CHART FOR ANIMIKIE AND PRE-ANIMIKIE ROCK UNITS IN THE LAKE SUPERIOR REGION*

	GUNFLINT RANGE	MESABI RANGE	CUYUNA RANGE	GOGEBIC RANGE	DICKINSON COUNTY		MARQUETTE RANGE	
	UPPER PRECAMBRIAN SEDIMENTARY AND IGNEOUS ROCKS (younger than 1.6 b.y.)							
	-------------------------------- -unconformity- --------------------------------							
Animikie Group	Rove Formation	Virginia Formation	Rabbit Lake Formation	Tyler Slate	Badwater Greenstone / Michigamme Slate / Hemlock Formation	Baraga Group	Michigamme Slate / Goodrich Quartzite	Marquette Range Supergroup
					-------------------- -disconformity- ------			
	Gunflint Iron-formation	Biwabik Iron-formation	Trommald Formation	Ironwood Iron-formation	Vulcan Iron-formation / Felch Formation	Menominee Group	Nagaunee Iron-formation	
	"Kakabeka Quartzite"	Pokegama Quartzite	Mahnomen Formation	Palms Quartzite			Siamo Slate / Ajibik Quartzite	
	------------------------------ -unconformity- ------------------------------							
			Trout Lake Formation quartzite and slate?	Bad River Dolomite		Chocolay Group	Wewe Slate	
				Sunday Quartzite	Randville Dolomite		Kona Dolomite / Mesnard Quartzite	
					Sturgeon Quartzite / Fern Creek Formation		Enchantment Lake Formation	
	------------------------------ -unconformity- ------------------------------							
	LOWER PRECAMBRIAN IGNEOUS AND METAMORPHIC ROCKS (older than 2.6 b.y.)							

*From Morey, 1972a and 1972b.

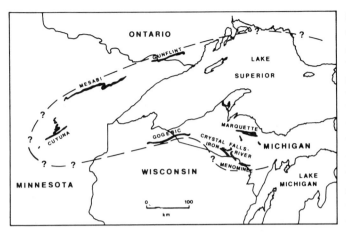

Figure 1. Generalized geologic map showing location of iron ranges (black) of the Lake Superior region. Dashed line shows an approximated boundary of the Animikie basin. The rocks to the north and west of the basin are Archean, the rocks to the south and east of the basin are Archean and Lower Proterozoic, and the rocks within the basin are Lower Proterozoic except for the Middle Proterozoic Keweenawan volcanics and associated rocks along the Mid-continent rift zone.

partly correlative with the lower portions of the Marquette Range Supergroup.

The long accepted correlations of the formations within the Animikie Group on the four ranges in the western portion of the basin are shown in Table 1. This paper deals primarily with only two of the basal units, the Palms Formation on the Gogebic range and the Pokegama Quartzite on the Mesabi range. Both were deposited as mud, silt, and sand sequences immediately beneath major iron-formations. The original sedimentary rocks have been transformed, probably by both burial metamorphism and low-grade (greenschist) regional metamorphism, into argillite, metasiltstone, and quartzite. The interpretation of the environments in which these clastic units were deposited is significant to a better understanding of the sedimentological regime in which the overlying iron-formations of the Lake Superior region were deposited.

The general outlines of the Animikie basin as described in the previous paragraphs are largely based on the present distributions of the rock units under discussion in this paper. Admittedly, the basin could have been considerably larger. For example, it is plausible that the basin was open to the south as depicted by Dott and Batten (1971, p. 170) and that what are herein interpreted as synchronous deposits on the opposite sides of a narrow basin could have slightly different ages as a result of deposition as different portions of transgressive sedimentary units. Thickness data are available across the depositional strike for some units along the Mesabi range and suggest proximity to a northern basin margin.

Unfortunately, such data are not available for the more steeply dipping units of the Gogebic range, so the

suggested proximity of these sedimentary units to a southern basin margin cannot be verified. The presence of large areas of Archean rocks to the south of the Proterozoic units, with no Proterozoic remnants on them, is interpreted as circumstantial evidence that an older landmass existed to the south of at least this part (an embayment?) of the larger basin which was broader to the east. Even if the basin had been open to the south, by Tyler time (Table 1), it was bounded on the south by a landmass, as determined by paleocurrent indicators which show that currents moved towards the northwest in the Tyler turbidite sequence (Alwin, 1979; Larue and Sloss, 1980). Also, the position of the basin astride the Great Lakes Tectonic Zone is suggestive of an elongate, confined zone of structural weakness. Fortunately, even if the geometry of the paleogeographic setting utilized here were found on the basis of new data to be incorrect, the sedimentologic interpretations of this paper would not be negated.

THE PALMS FORMATION

The Palms was named by Van Hise (1901); and it has been described by Irving and Van Hise (1892), Van Hise (1901), Van Hise and Leith (1911), Hotchkiss (1919), Aldrich (1929), Leith, Lund and Leith (1935), Tyler and Twenhofel (1952), and Prinz (1967). Schmidt (1980) presented an excellent summary of past work and added numerous new observations. Larue (1981a) did an extensive sedimentological study in the southern Lake Superior region and included interpretations of the Animikie sequence.

Distribution and Thickness

The Palms Formation is exposed at numerous localities along the length of the Gogebic range and a few tens of kilometers further east as well, for a total distance of about 90 km. The formation has a maximum measured thickness of about 150 m although it may be thicker elsewhere on the range. It dips at about 65° to the north-northwest at most localities.

Contact Relationships

The Palms Formation unconformably overlies Archean granitic and volcanic rocks and locally the Lower Proterozoic Bad River Dolomite.

The contact between the Palms and the overlying Ironwood Iron Formation is well exposed at one locality in Wisconsin where it appears to be gradational over less than one meter of strata. However, thin lenses of sandstone (quartzite) are present in the overlying cherty iron-formation as much as 5 m above the contact; and at another locality, iron-rich beds a few cm thick occur in the upper middle portion of the Palms Formation.

General Lithology

The Palms Formation contains three major rock types: quartzite, metasiltstone, and argillite. Throughout much of this paper, the premetamorphic lithologic terms "sandstone, siltstone, and shale" will be used. The formation has long been described as consisting of two main members, a thick lower argillite ("quartz-slate") member and a thinner upper quartzite member. Some early workers (e.g., Van Hise and Leith, 1911; Aldrich, 1929) considered local beds of basal conglomerate as a third member. These schemes are suitable for general field identification, but for a more detailed sedimentologic analysis, a sedimentary facies approach is desirable.

Four facies will be described: a basal conglomerate facies, a shale facies, a heterolithic shale-siltstone-sandstone facies, and a sandstone facies. The relationships of these facies are evident in the columnar sections shown in Figure 2. Where the basal conglomerate facies is missing, the shale facies directly overlies older rocks. The shale-siltstone-sandstone facies dominates the formation, and the sandstone facies is invariably the uppermost facies, but is locally present lower in the column as well.

Conglomerate facies. The conglomerate facies is the least abundant of the sedimentary facies in the Palms. Where present, it occurs as thin lenses overlying older rocks. It consists largely of clasts derived from underlying bedrock units plus clasts of quartz, chert, and jasper. At one Michigan locality, a phosphatic mudchip and chert fragment conglomerate 15 cm thick is present at the contact with silicified Bad River Dolomite. A few meters of basal granite-cobble conglomerate are present above Archean granite at other localities in Wisconsin. Aldrich (1929) and others have described additional occurrences.

Pebble roundness and size vary with the compositions of the basement rocks. Chert clasts are only a few cm in diameter and are angular, whereas greenstone or granite clasts may be as large as 15 cm in diameter and are subangular to rounded. All the observed conglomerates are massive and clast-supported with only interstitial sandy matrix and without sandy interbeds.

Shale facies. The shale facies is the basal unit of the Palms except where conglomerate is present (Fig. 2), but it also occurs interbedded with the shale-siltstone-sandstone facies at a few localities as units 0.5 to 2 m thick.

Although shale dominates in this facies, thin laminae of siltstone and beds of fine-grained sandstone also are commonly present. Essentially all bedding is parallel, with laminae generally on the order of 2 to 5 mm thick (Fig. 3).

Small-scale mudcracks were observed within this facies at one locality, but what appear to be nontectonic wrinkle marks (runzel marks) are common. Minor miscellaneous bed markings, commonly sole marks, of the types described and illustrated below in the shale-siltstone-sandstone facies also are present.

Shale-siltstone-sandstone facies. This is the dominant facies in the Palms Formation (Fig. 2) and includes a great variety of sedimentary structures. The three rock types alternate in a more-or-less random fashion, varying from ap-

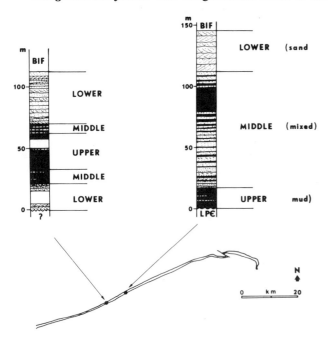

Figure 2. Two measured sections of the Palms Formation on the Gogebic range. Dark shading represents shale, dotted patterns represent siltstone and sandstone, blank spaces represent covered intervals, BIF is banded iron-formation, and LPЄ is Lower Precambrian (Archean) basement rock. The terms "lower, middle, and upper" refer to tidal facies. Sections are on Tyler Fork and Potato River in Wisconsin.

Figure 3. Shale facies at base of Palms Formation. Location is Newport Mine near Bessemer, Michigan.

Figure 4. Parallel, wavy, and lenticular beds in shale-siltstone-sandstone facies of the Palms Formation. Location is Radio Tower Hill in Wakefield, Michigan.

Figure 5. Thin and thick parallel, wavy, and lenticular beds of shale-siltstone-sandstone facies of the Palms Formation. Location is on Montreal River at Gile, Wisconsin.

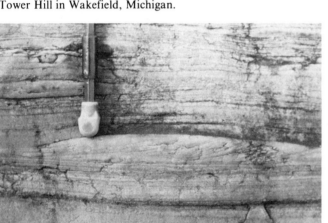

Figure 6. Minor low-relief scour at top of photo and cross-bedded sand lens below pencil in the shale-siltstone-sandstone facies of the Palms Formation. Location is on Tyler Fork, Wisconsin.

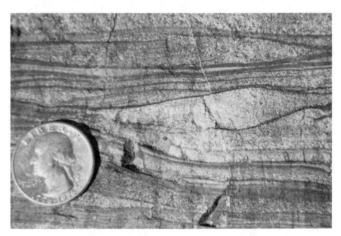

Figure 7. Thin muddy flaser beds between thin beds and between cross-beds in the shale-siltstone-sandstone facies of the Palms Formation. Coin is 24 mm in diameter. Location is Radio Tower Hill in Wakefield, Michigan.

proximately equal amounts to each being dominant in short sequences.

The most obvious feature of the facies is its thin-bedded nature. Four main types of bedding are present. Thin parallel to wavy beds are dominant (Fig. 4), with rare beds as thick as 0.5 meters (Fig. 5). Lenticular beds of sandstone, commonly cross-bedded, are common in sequences dominated by shale and siltstone and vary from 10 cm to several meters in length. Most are 1 to 6 cm thick, but some are as much as 15 cm thick (Fig. 6). Where siltstone or sandstone are dominant, thin flaser beds (mud drapes) a few mm thick occur between beds or on cross-bed laminae (Fig. 7). Graded bedding (Fig. 8) is rare.

Several other sedimentary structures also are present. Low relief scour features occur locally (Fig. 6). Abundant current marks are commonly (typically) miscellaneous drag (?) marks of a myriad of types; flute casts are rare (Fig. 9).

Herringbone cross-beds and reactivation surfaces are uncommon (Fig. 10). Also uncommon are thin zones of soft-sediment deformation.

Sandstone facies. The shale-siltstone-sandstone facies becomes sandier upward and passes gradually into the sandstone facies characterized by thick sandstone beds and a general absence of shale and siltstone. The sandstone beds become thicker upward; in the lower part of the facies, beds are as much as 15 cm thick whereas higher in the facies, they are as much as one meter thick (Fig. 11).

A common feature is tabular (planar) cross-bedding. Ripple marks, generally symmetrical, are common in the lower part of the unit; and flat-topped ripples were noted on one bedding surface low in the unit. Most ripple marks are broad and low, with ripple indices of 14 to 30 and ripple symmetry indices (Reineck and Singh, 1975) of 1 to 4. At one locality low in the facies, mudcracks and sand lags are

Figure 8. Graded sandstone bed in the shale-siltstone-sandstone facies of the Palms Formation. Coin is 18 mm in diameter. Location is on Tyler Fork in Wisconsin.

Figure 9. Flute casts on sole of fine-grained sandstone bed in the shale-siltstone-sandstone facies of the Palms Formation. Location is east of Wakefield, Michigan.

Figure 10. Herringbone cross-bedding in shale-siltstone-sandstone facies of the Palms Formation. Coin is 24 mm in diameter. Location is Radio Tower Hill in Wakefield, Michigan.

Figure 11. Sandstone facies of the Upper Palms Formation. Staff is about 1.5 m long. Location is on Tyler Fork, Wisconsin.

present. However, thorough cementation by silica has resulted in a well-indurated rock with a vitreous luster; sedimentary structures are generally obscured.

General Petrography

A detailed petrographic study is beyond the scope of this paper. The framework grains of the sandstones are mostly rounded monocrystalline quartz, with minor feldspar, chert, and iron silicates; some sandstone units may be classified as quartz arenites (Fig. 12), others as feldspathic arenites, and many as quartz wackes with appreciable clayey matrix. The sandstones of the sandstone facies are generally more mature, texturally and mineralogically, than the thin-bedded sandstones of the shale-siltstone-sandstone facies (Fig. 13). Several heavy mineral mounts show that rounded zircon and less abundant rounded tourmaline dominate the nonopaque heavy mineral suite.

Larue (1981a) has done some detailed petrography and grain-size determinations; the mean grain-size for the sandstones ranges from 0.1 to 1.2 mm, with most at about 0.4 mm.

POKEGAMA QUARTZITE

The Pokegama Quartzite has been studied briefly by various workers since it was named by H. V. Winchell (in Winchell, N. H., 1893) for exposures at the west end of the Mesabi range. Much of the previous work has been summarized by Morey (1972b; 1973). The most notable older reports are those of Leith (1903), Grout and others (1951), and Dolence (1961), who did the most complete work.

Distribution and Thickness

Few natural exposures exist, for thick glacial drift gen-

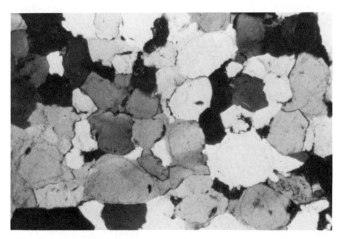

Figure 12. Photomicrograph of silica-cemented quartz sandstone of the sandstone facies of the Palms Formation. Field of view is 1.5 mm across. Sample is from Radio Tower Hill in Wakefield, Michigan.

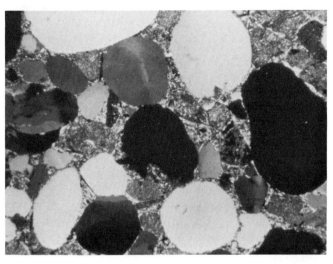

Figure 13. Sericitic quartz sandstone of the shale-siltstone-sandstone facies of the Palms Formation. Field of view is 2.5 mm across. Sample is from Ramsey, Michigan.

erally covers the formation. Outcrops, roadcuts, and mine cuts occur at a few places over the length of the Mesabi range, a distance of about 130 km; but most exposures are restricted to the central part of the range. Only two drill holes, one located just south of Eveleth (NE¼, NE¼, Sec. 5, T. 57N., R. 17W.) and the other southwest of Mountain Iron (SE¼, SE¼, Sec. 8, T. 58 N., R. 18W.), separated by 12 km, are known to transect the entire formation which is 51 m thick and 26 m thick, respectively, at these two points (Fig. 14). Several other drill holes have penetrated the upper part of the formation, but most drilling on the Mesabi range was performed in relation to iron ore exploration and development and barely penetrated the Pokegama beneath the Biwabik Iron Formation. The Pokegama is thought to thicken from the eastern end of the Mesabi range, where it is only a few meters thick, to the western end of the range where it may be as much as 100 m thick (White, 1954, p. 20; Morey, 1979). It also thickens to the south in the central part of the range.

Contact Relationships

The Pokegama Quartzite unconformably overlies Archean metavolcanic and metasedimentary rocks. Gray chert and jasper precipitated in cracks on this Archean surface have yielded specimens of microscopic life forms (Gruner, 1922, 1923; Cloud and Licari, 1972); the chert and jasper have been interpreted as Pokegama by several workers, but could have been directly overlain by iron-formation and hence could be parts of the Biwabik Iron Formation. There may be as much as 30 m of relief on the Archean surface beneath the quartzite (Grout and Broderick, 1919), but the surface was, nevertheless, essentially a peneplain.

The contact of the Pokegama and the overlying Biwabik Iron Formation is conformable, although a chert-fragment conglomerate at the contact has been interpreted as a minor unconformity (Leith, 1903, p. 167). Dolence (1961) described both gradational and sharp contacts. A cherty zone found 6 m and 12 m beneath the top of the Pokegama in two different drill cores (Dolence, 1961) further signifies a gradation between the Pokegama and the Biwabik (Fig. 14).

General Lithology

The Pokegama Quartzite, in spite of its name, consists of three main rock types: quartzite, metasiltstone, and argil-

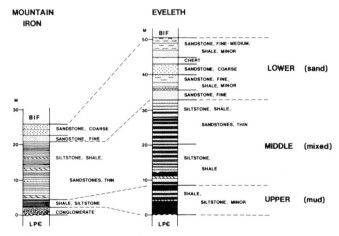

Figure 14. Measured sections from drill cores of the Pokegama Quartzite. Dark shading represents shale, thin blank units represent siltstone, dotted patterns represent sandstone, BIF is banded iron-formation, and LPЄ is Lower Precambrian (Archean) basement. The terms "lower, middle, and upper" refer to tidal facies. Eveleth log after Dolence, 1961. Highly generalized.

Figure 15. Thin parallel beds and cross-beds in the shale-siltstone-sandstone facies of the Pokegama Quartzite. Coin is 24 mm in diameter. Location is Eveleth, Minnesota.

Figure 16. Current marks on sole of a thin sandstone bed in the shale-siltstone-sandstone facies of the Pokegama Quartzite. Coin is 24 mm in diameter. Location is Eveleth, Minnesota.

lite. Minor basal conglomerate is present locally. Premetamorphic lithologic terms will generally be used in the following descriptions.

Most exposures consist of massive quartzite that occurs just beneath the Biwabik Iron Formation. However, a few artificial exposures of lower parts of the formation are made up largely of thin-bedded argillite and metasiltstone. The drill holes which cut the entire formation show best the stratigraphy of the formation (Fig. 14). The top part of the formation consists of sandstone, mostly fine- to medium-grained with thin siltstone and shale layers; the middle portion consists of interbedded siltstone and shale with only very minor sandstone; and the lowest consists of shale with minor siltstone.

The same four facies used in describing the Palms—conglomerate, shale, shale-siltstone-sandstone, and sandstone—will be utilized for the Pokegama. The vertical relationships of the facies are illustrated in Figure 14.

Conglomerate facies. Small remnants of a basal conglomerate are exposed on the flanks of metagraywacke knobs near the central part of the range. Many clasts consist of white vein quartz, but most clasts match the underlying Archean bedrock, as also noted by Leith (1903, p. 98). In the Mountain Iron drill hole, rounded clasts of Archean metasedimentary rocks, as large as 3 cm in diameter, make up the two meters of conglomerate.

Shale facies. The shale facies can be studied in only one roadcut and in the two cores that have penetrated the entire formation (Fig. 14). It consists dominantly of thinly bedded shale, but some silt laminae and sandstone beds also are present. Minor channeling is common at the base of the thicker siltstone and sandstone beds, and at one locality, 0.5 m of section has been eroded. Small-scale cross-bedding occurs in some siltstone beds, as do soft-

sediment deformation structures. Concretions as large as 15 cm also are present.

Shale-siltstone-sandstone-facies. This facies is dominated by thin-bedded siltstone with minor shale and sandstone. It is exposed at only two localities in the central part of the range, but the drill holes shown in Figure 14 consist largely of this facies.

Much of the facies is either parallel- or wavy-bedded (Fig. 15). Small-scale cross-bedding is common. Flaser bedding also occurs as very thin clay partings. Small symmetrical ripple marks are common. Parting lineations and current marks of various types, including small flute casts (Fig. 16), and runzel marks are abundant. Small circular structures which are probably either raindrop imprints or degassing structures are rare. Troughs as large as 7 cm deep by 20 cm wide also are present and provide valuable paleocurrent data. Minor scour features also occur beneath some thin sandstone beds.

Sandstone facies. The sandstone facies is characterized by beds of sandstone as much as 1.5 m thick, separated by thin beds of shale or siltstone which weather as deep reentrants on outcrops. The resistant sandstone beds appear massive, but many contain some cross-beds which have been partially obscured by thorough silica cementation. A few ripple marks have been observed and possible syneresis cracks are also present. Some beds are coarse-grained enough to be called grit, but most consist of medium- to coarse-grained sand.

General Petrography

The general statements on the petrography of the Palms Formation apply to the Pokegama Quartzite as well.

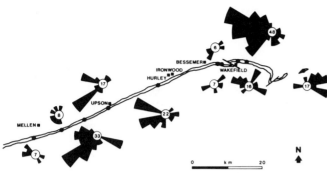

Figure 17. Map of Gogebic range showing cross-bedding plots in the Palms Formation at 10 locations. Note bimodal-bipolar pattern at several localities.

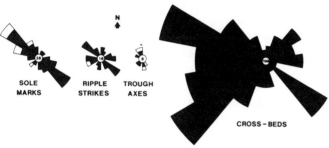

Figure 18. Total paleocurrent measurements in the Palms Formation. Note overall bimodal-bipolar aspect of the cross-bed plot. White portion of noncross-bed plots are sense measurements.

PALEOCURRENT ANALYSIS

Figure 17 illustrates the paleocurrent plots for several individual localities in the Palms Formation. A total of 199 cross-beds was measured (Fig. 18); each was rotated on a stereonet about the strike direction to its original pretectonic orientation, assuming in most cases only a simple tilt to the north during deformation. This assumption seems valid, based on the long strike length of the formation and the nearly total lack of folds except at one locality where they may be a product of drag along a major fault. In addition, 8 trough axes, 28 sole marks, and 16 ripple marks were measured (Fig. 18). Several of the individual cross-bedding plots are bimodal-bipolar, with major modes trending generally to the west and east. The total cross-bedding plot also shows a similar pattern as well as a less pronounced north-south bimodal-bipolar trend. Vertical plots of paleocurrents through one representative measured section show that the currents oscillated back and forth (Fig. 19).

Because of the lack of Pokegama exposures, only 38 paleocurrent indicators were measured. These show a crude bimodal-bipolar pattern with modes to the north and south (Fig. 20).

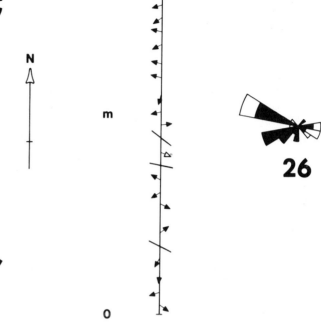

Figure 19. Vertical variation in paleocurrent directions within a measured section of the Palms Formation on the Montreal River at Gile, Wisconsin. Arrows represent azimuths on a compass. Readings are in order but are not plotted at actual locations within column. The black arrows represent cross-beds, the white arrow represents a flute cast, and the black lines represent bidirectional sole marks.

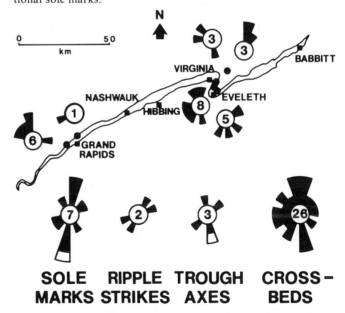

Figure 20. Map of Mesabi range showing cross-bedding plots in the Pokegama Quartzite at 6 localities. Total cross-bedding plot is at lower right. Also shown are plots of sole marks, ripples, and troughs; the white measurements represent sense measurements.

PALEOENVIRONMENTAL INTERPRETATION OF THE PALMS FORMATION AND THE POKEGAMA QUARTZITE

A tidally-influenced transgressive marine model fits the data from both the Palms Formation and the Pokegama Quartzite. The case for the better exposed Palms is particularly strong, with the bimodal-bipolar paleocurrent patterns, the nature of the bedding, other sedimentary structures, the facies relationships, and the mineralogical maturity of the original sediment, all supporting the tidal model. The model is less well established for the Pokegama, partly because of the lack of exposures, but nevertheless the available data are compatible with the model. The Pokegama sediments were probably accumulating along the northern margin of a 75-100 km wide embayment of a transgressive sea at about the same time as sediments of the Palms were being deposited along the southern margin of the embayment. The basin was not starved; sediment was supplied from the adjacent low-lying landmasses at a slow rate.

The sedimentary structures—including the presence of alternating thin beds of shale, siltstone, and sandstone and the lenticular, parallel, wavy, and flaser bedding types—correspond well with those associated with a tidal environment as detailed by numerous workers including de Raaf and Boersma (1971), Reineck and Singh (1975), Klein (1977b), Johnson (1978), and Elliott (1978). The structures are remarkably similar to those in the Archean Moodies Group of South Africa which Eriksson (1977) interpreted as having formed in a tidal environment. The coarse sediments appear to have been transported by high-energy tidal and storm currents via bedload traction, whereas the fine sediments appear to have settled out of suspension during the slack high-water stages of the tidal cycle and during calms after wave-induced turbulence (e.g., McCave, 1970; Anderton, 1976).

A low tidal range and a microtidal coastline (Blatt and others, 1980) are indicated by a lack of channels (e.g., Reineck and Singh, 1975, p. 356). Only a few low-relief channels were observed in the two formations. Even in the Wash of England, which has a tidal range of as much as 6.7 m, Evans (1965) has shown that tidal channels are of little importance. Klein (1971a; 1972) noted that paleotidal ranges are best documented on coastlines characterized by moderate to rapid rates of progradation; the opposite situation, one of a major transgression, apparently dominated during Palms and Pokegama sedimentation. Minor progradations, however, are indicated within the rock column by fining-upward sequence, and at one locality a fining-upward sequence of shale to sandstone may suggest a possible paleotidal range of 7.5 m. In general, the lack of evidence for a large paleotidal range suggests that the tidally-influenced shelves were narrow (Klein, 1977a; Klein

and Ryer, 1978), although Komar (1976) noted that shelf width and tidal range along embayed coasts may not be directly related.

Whereas burrowing is an aid in environmental determinations in Phanerozoic tidal sequences (e.g., Thompson, 1968; Yeo and Risk, 1981), the absence of bioturbation in these Precambrian sediments has helped to preserve the thin bedding.

The following interpretations of the significance of the four facies found in the two formations provide additional details of the model. Because the facies in the Palms and Pokegama are so similar, they will be interpreted together.

Interpretation of the conglomerate facies. The conglomerate facies was present prior to the transgression of the Animikie Sea, occupying small, low-relief depressions on a peneplaned surface, most of which had undergone erosion for the previous 600 million years or so. Residual weathering and minor fluvial action were probably both instrumental in the initial accumulation of the discontinuous, clast-supported conglomerates.

The basal conglomerate facies of the Pokegama Quartzite in drill cores and small outcrops consists of subrounded to rounded cobbles and boulders of the underlying Archean bedrock and vein quartz. Reworking by the transgressing sea seems likely, for some remnants are plastered on the flanks of recently exhumed, small monadnocks of Archean rock which must have been near-shore islands in the Animikie Sea.

The presence of apparently locally derived phosphatic mudchips in the conglomerate facies at the unconformity between the Palms and underlying Bad River Dolomite, and between the Palms and Archean granite as well, lends support for a marine origin of the lowermost Palms sediments at some localities (see Johnson, 1978, p. 230) and for long exposure of the surface to sea water with essentially no sedimentation. At other localities, thicker basal conglomerates may be fluvial accumulations in slight depressions.

Interpretation of the shale facies. The shale facies fits the suspensional depositional tidal phase of Klein (1977b), with deposition of suspended material during high-water slack tide periods. Much of the mud may have been placed in suspension by storms, with the cloudy water moved onto the upper tidal flat by both tidal currents and wave action. Most of the silt and clay comprising the mud settled out, perhaps with slight reworking, as parallel laminated beds of silt and mud. Accumulations of mud on the upper tidal flat, especially in areas with low-lying coastlines, is common along modern coasts in northwestern Europe (e.g., Evans, 1965; Reineck and Singh, 1975) and in eastern North America (e.g., Klein, 1963).

Sedimentary structures theoretically characteristic of the upper tidal flat, where very shallow water and periodic exposure and dessication might be expected, include runzel marks which may have numerous origins including the wrinkling by wind of sediment surfaces beneath thin-water films (see Reineck and Singh, 1975, p. 56; Klein, 1977b), mudcracks, and beds with abundant mudchips. Mudcracks are rare in the formations under study here, and although this is disconcerting, their rarity may be explained by a climate that prevented dessication (see Van Straaten, 1959; and Reineck and Singh, 1975); furthermore, such intertidal flats may not be exposed long enough for dessication, which would be more prevalent on supratidal, rarely wetted flats. Johnson (1975) and Watchorn (1980) have noted a similar lack of mudcracks in sequences which bear other characteristics of an upper tidal flat origin. The lack of mudchip layers can be explained by the above reasoning, by a poor preservation potential (Klein, 1971b), or by the fact that mudchip conglomerates are best developed in channels (Reineck and Singh, 1975, p. 356) which are uncommon in these units.

Interpretation of the shale-siltstone-sandstone facies. This facies, which comprises most of the Palms Formation and the Pokegama Quartzite, is the product of an alternation of bedload and suspension depositional processes. Associated interbeds of mud, silt, and sand and wavy beds, parallel beds, lenticular beds, and flaser beds are most common in "mixed" intertidal flats, between low- and high-tide marks (Reineck and Wunderlich, 1968; Reineck, 1972; Reineck and Singh, 1975; Klein, 1977b). This so-called "tidal bedding" (Reineck and Wunderlich, 1968; Reineck, 1972) is probably the most distinctive characteristic of a tidal flat environment. The abundance of the bedding types varies somewhat with the texture of the sediment; where sand and silt beds are dominant, minor flaser beds of mud are present, the product of settling from slack water. Where mud dominates, lenticular beds of sand or silt, which formed as isolated starved ripples that migrated across a cohesive muddy bottom, are present. Wavy beds are present where both sand and mud are preserved, with mud deposited in the ripple troughs of the underlying sand (Reineck and Singh, 1975). Parallel beds may be interpreted as the product of upper flow regimes, of plane bed transport at low velocities (Klein, 1977b, p. 18), or of sedimentation from suspension clouds in an environment with an energy too low to form ripples (Reineck and Singh, 1972; 1975). Low velocity origins seem the most likely, but the presence of parting lineation in the Ppokegama indicates some deposition in an upper flow regime, and minor scour features verify occasional appreciable velocities.

Tidal currents locally possessed sufficiently high velocities to rework and deposit the sand and silt as ripples and as larger-scale cross-beds. Herringbone cross-beds are sparse, but Reineck and Singh (1975) noted that they are best developed in channels which are uncommon in these units. Rare graded beds may have settled from storm-generated suspension clouds (Johnson, 1978, p. 233). Various current marks, including sole marks, are most probably the result of tools dragged across the fine-grained substrate, but some are the products of scour by currents. Runzel marks, mudcracks, and raindrop impressions or degassing structures indicate subaerial or near-subaerial exposure, and symmetrical ripple marks suggest shallow water.

Interpretation of the sandstone facies. The sandstone facies consists almost entirely of sand, signifying a high-energy environment. The beds, as thick as 1.5 m, generally are parallel laminated or planar cross-stratified; a few thick beds are clearly an amalgamation of two or more subunits which represent two or more depositional events.

A lower tidal flat environment near the low-water line or a shallow subtidal environment strongly influenced by tidal currents is the environment which best fits a unified model for this sedimentary facies (see Johnson, 1977). The sands probably accumulated as lower tidal sand flats or as subtidal sand shoals in water above wave base, for symmetrical ripple marks are abundant. Locally, the shoals built up to the water surface and became emergent, as indicated at one locality by flat-topped ripples similar to those described by Brunn and Mason (1977) and Beukes (1977), and by mudcracks. Syneresis cracks are compatible with very shallow water. Coarse sand lags indicate minor winnowing after deposition. Perhaps some shoals even became barrier islands, as might be expected in a microtidal environment (Blatt et al., 1980, p. 656). In a modern example, barrier islands replace shoals west of the Elbe estuary in Germany where the tidal range is low (Reineck and Singh, 1975).

An alternative to a lower tidal or subtidal, marine origin for the sandstone facies is deposition on a beach. Arguing against such an origin is the relatively high average amount of dip of the cross-beds, about 17°, for both the Palms and the Pokegama; beach cross-beds generally dip at angles of less than 10° (Pettijohn et al., 1973, p. 480).

Facies relationships. The overall vertical relationships of the four facies described here in both the Palms and Pokegama formations are well established (Figs. 2 and 14). However, the outcrops of steeply dipping Palms Formation present only a narrow, elongate, two-dimensional slice of the formation; and the Pokegama outcrops are so few and scattered that they hardly provide for any three-dimensional data. Nevertheless, by applying the familiar Walther's Law of Succession of Facies, which essentially states that "facies sequences observed vertically are also found laterally" (Blatt et al., 1980), the lateral relationships can be interpreted as follows. The conglomerate facies was locally present on the low-lying peneplaned surface prior to transgression of the Animikie Sea. The upper tidal flat (shale) facies was the most proximal marine facies resting upon conglomerate or bedrock. Seaward of the upper tidal flat was the middle tidal flat (shale-siltstone-sandstone) facies and then further seaward, the lower tidal flat and subtidal (sandstone) facies.

Ignoring the conglomerate, which probably has a non-tidal origin, the Palms and Pokegama are coarsening-upward sequences which are here interpreted to be the products of a marine transgression, with the finer sediments deposited near the high-water mark and the coarser sediments deposited near the low-water mark. Such a model was described by Reineck (1972) and others, and a modern transgressive tidal model has been presented by Yeo and Risk (1981). Most ancient tidal sequences described in the literature, however, fine upward and are interpreted to be the result of prograding tidal environments. There are fining-upward sequences (see Fig. 2) within some sections of the Palms Formation that indicate minor progressions, but the overall pattern is one of a major transgression.

An integral facet of the tidal facies interpretation is the paleocurrent information. Most of the cross-bed measurements in both the Palms and the Pokegama are from the middle tidal flat facies. A general bimodal-bipolar overall pattern emerges for the Palms, with minor modes giving it a quadrimodal aspect as well (Fig. 18); such patterns are described as characteristic of intertidal flats by Klein (1977b, p. 71) and others. The dominant mode is here interpreted to be the result of flood tides, rather than the more typical situation where ebb tides are dominant (e.g., Reineck and Singh, 1975, p. 320). The major paleocurrent modes in the Palms are parallel to the probable shoreline, whether or not an embayment was present, as discussed in the introduction. This relationship has been noted commonly in modern tidal environments (e.g., Evans, 1965, in the Wash of England; Klein, 1968, in the Minas Basin of Nova Scotia) and in studies of ancient tidal environments (e.g., Tankard and Hobday, 1977, in the Ordovician of South Africa; and Klein and Ryer, 1978, for various aged rocks). Klein (1977b) emphasized that this parallelism is the general case and is consistent with a tidal basin circulation model. This parallelism could be used as evidence for an embayment type of elongated tidal basin, but such reasoning may be too circular. At one locality in the Palms Formation where a number of readings were obtained from the subtidal or lower tidal (sandstone) facies, the pattern is unimodal to the west-southwest, parallel to the interpreted flood tide direction.

Also consistent with a tidal environment of deposition for the sediments is the textural and mineralogic maturity of the sand. A high degree of sand grain roundness is prominent, as described in tidal environments by Swett et al. (1971) and Balazs and Klein (1972). The best rounding and sorting appears to be found in the sandstone facies where energy was highest and where the reworking time would have been longest (Fig. 12). However, the roundness of the dominant quartz sand grains cannot be attributed only to maturation in a tidal environment. Some recycling from the underlying Sunday Quartzite is likely; and even first cycle sand eroded off the vegetation-free, well-weathered, and undoubtedly wind-swept adjacent peneplain may have already consisted largely of eolian-rounded quartz grains. Nevertheless, at least a moderate amount of residual time in a tide-dominated environment seems likely in this tidal model, with the sand apparently moved shoreward from slightly deeper water by storms and tidal currents as noted by De Jong (1977) on Dutch tidal flats.

A tidal interpretation does not imply that the relatively thin Palms and Pokegama formations represent the total thickness of all material ever deposited in the tidal environment, nor that all thin laminae and beds are simply the result of diurnal tides. The formations undoubtedly represent a long but indeterminate period of tidal deposition interspersed with storms of unknown frequency and with intervals of little or no deposition or even erosion. The environment appears to have been tide-dominated.

Most previous workers have concluded that the Pokegama was deposited in a shallow marine environment (e.g., White, 1954; Dolence, 1961; Cloud and Licari, 1972; Morey, 1973). Deposition in restricted back arc basins was proposed by Van Schmus (1976). However, Tyler and Twenhofel (1952, p. 137) preferred deposition in a fresh water basin.

Simonson (1982) concluded, largely on the basis of sedimentary structures, that both the Palms and the Pokegama formations were deposited as shelf sands, which may have been in part lower intertidal. The Palms has been interpreted as a transgressive, platformal shallow marine sequence by Larue and Sloss (1980) and especially by Larue (1981a). He also concluded that the shale-siltstone-sandstone facies of this study was deposited in a wave-dominated, offshore shelf environment, a conclusion which is at odds with the model developed in this paper. Reasons for preferring the present model to one such as Larue's are listed in the Summary and Conclusions section.

SIGNIFICANCE OF THE TIDAL MODEL TO THE OVERLYING BANDED IRON-FORMATION

The Palms Formation and the Pokegama Quartzite have fairly sharp but nevertheless gradational boundaries with the overlying Ironwood and Biwabik Iron Formations, both of which are more than 200 m thick (Table 1). Therefore, Walther's Law of Succession of Facies can again be utilized, this time to ascertain the sites of deposition of the iron-formations. Such reasoning places the sites on the shelf just seaward of the lower tidal or subtidal (sandstone) facies, away from the abundant clastics nearer shore which would have masked any iron minerals that might have been precipitating there. Similarly, minor clay that settled in the regime of abundant iron mineral precipitation would have been masked by iron-formation; this may be reflected by the low alumina content of the iron-formation, which averages 1.0 percent in the Biwabik (White, 1954) and 1.7 percent in the Ironwood (Schmidt, 1980). Winnowing of fine-grained material from shallow-water, higher-energy environments and deposition in deeper-water, lower-energy environments may have occurred.

Various detailed lines of evidence show that the tidal terrigenous deposits and the overlying iron-formations are genetically related. Within the inferred intertidal facies of the Palms at one locality, granules of iron-silicates comprise a few beds less than 4 cm thick, and may have been transported shoreward by tidal currents and/or storm surges. Also, thin cherty layers have been observed high in the Palms (Schmidt, 1980). In the lower tidal or subtidal (sandstone) facies of the Pokegama (Fig. 14), there is a unit nearly 2 m thick consisting of ferruginous sand-sized chert grains probably derived from the site of iron-formation deposition, which in this model would have been situated seaward of the sandstone facies. Conversely, sandstone lenses occur in the Ironwood Iron Formation several meters above the base of the formation, and Aldrich (1929) noted quartz grains suspended in chert in the lower 3 m of the Ironwood. Near the west end of the Gogebic range, a thin layer of quartz sand was noted in drill core near the middle of the Ironwood (oral communication, R. W. Marsden, 1981). White (1954, p. 15-16) described terrigenous clastic units within the Biwabik.

White (1954) concluded after a detailed study of the Mesabi range that the Biwabik Iron Formation was genetically related to a shoreline and was closely associated with contemporaneous clastic sediments of a normal marine character. White further suggested that the two main lithofacies of the Biwabik Iron Formation (a coarser-grained, thicker-bedded, iron oxide-chert "cherty" facies and a finer-grained, thinner-bedded iron silicate-iron carbonate "slaty" facies) are depth-related, with the cherty facies being a shallower-water, detrital deposit reworked by waves and currents. These two facies are each repeated, making up four stacked members named, from bottom to top, the Lower Cherty, Lower Slaty, Upper Cherty, and Upper Slaty (Wolff, 1917), with minor interbedding of the two facies in each unit. White proposed that this major repetition was due to two transgressions separated by a regression. White's conclusions mesh well with the data gathered in this study; the initial major transgression was followed by a minor regression and a minor transgression.

A similar relationship of "cherty" and "slaty" facies has long been known in the Ironwood Iron Formation and has recently been described in detail by Schmidt (1980). The sequence is, from the base upward, the Plymouth Member ("cherty"), the Yale Member ("slaty"), the Norrie Member ("cherty"), the Pence Member ("slaty"), and the Anvil Member ("cherty"). Aldrich (1929) noted 12 "cherty" to "slaty" transitions in the Wisconsin portion of the Ironwood Iron Formation.

Several other workers have suggested general environments of deposition for the Lower Proterozoic banded iron-formations in the Lake Superior region, and most agree on a shallow marine origin (e.g., James, 1954; Gross, 1972, 1973, 1980; Mengel, 1973; Bayley and James, 1973); but there is a bewildering wide range of specific environments which have been proposed. Certainly, additional work needs to be done. James (1954) specified deposition of the Nagaunee and Vulcan Iron Formations of Michigan in individual restricted marine basins which formed as a result of broad offshore buckles marginal to the shorelines. Eugster and Chou (1973) called upon barred or partially barred lagoons or basins with wide supratidal flats. For iron-formations in the Transvaal Super-group of South Africa, Button (1976) has suggested restricted environments behind carbonate shoals. Drever (1974) suggested deposition along continental margins near upwelling iron- and silica-rich bottom waters. Chauvel and Dimroth (1974) and Dimroth (1979) delineated shelf-marginal banks and lagoonal platforms for the Sokoman Iron Formation of Labrador. Goodwin (1956) preferred a structural control of water depth as the reason for a similar facies pattern in the Gunflint Iron Formation. Shegelski (1980) interpreted some lithologies in the upper portion of the Gunflint to have formed in a littoral-tidal environment. However, Sakamato (1950) preferred a lacustrine setting. Simonson (1982) concluded, on the basis of a variety of sedimentological criteria in a number of iron-formations, that what he calls the "arenitic" type of iron-formation (the "cherty" of this paper) is a shallow-water deposit.

Huber (1959) placed the clastic ("cherty") portions of the Ironwood Iron Formation nearshore and the chemical ("slaty") facies in somewhat deeper water, and called upon partially restricted circulation to enhance variations in physical and chemical conditions and, consequently, in the iron-formation facies. Larue (1981a) preferred a continental margin (offshore shelf) site for the deposition of the Ironwood Iron Formation.

Evidence of a higher-energy, probably shallower-water, origin for the "cherty" facies of the Biwabik than for the "slaty" facies includes the following: (1) There are two algal chert horizons present, one at the base of the Lower Cherty Member near the base of the formation and the other near the middle of the Upper Cherty Member, both with detrital material in intermound areas. (2) Numerous chert fragment layers in the "cherty" facies appear to be the result of storm activity. (3) The chert and iron mineral grains in the "cherty" facies are rounded and appear to have been reworked. (4) Abundant cross-bedding is present in thicker-bedded portions of the Lower Cherty Member in the central part of the Mesabi range (Figs. 21 and 22) as well as in the western part (Mengel, 1973), but missing in interbedded "slaty" units which instead show well-preserved thin laminae characteristic of lower energy. Morey (1973) concurred with a shallower-water origin for the

Figure 21. Cross-bedded "cherty" Biwabik Iron Formation northeast of Virginia, Minnesota.

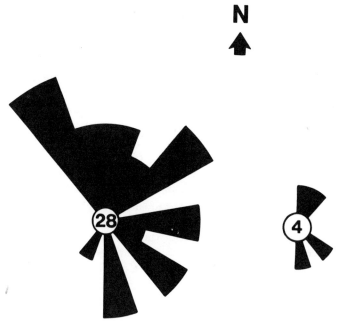

Figure 22. Plots of cross-bed measurements in "cherty" Biwabik Iron Formation at two localities, Eveleth and northeast of Virginia, Minnesota.

"cherty" facies. A number of south-trending channels in iron-formation, as long as 2 km, as wide as 800 m, and with reliefs of tens of meters, have been outlined by mining and drilling at several places on the Mesabi range (White, 1954), including U.S. Steel's Minntac pits at Mountain Iron (personal communication, Wayne Plummer, 1981). These channels are cut into the Lower Slaty but contain interbeds of cherty granular material similar to the Lower Cherty; one channel I observed contained cross-beds. It is likely that these are subtidal channels cut approximately perpendicular to the shoreline by tidal currents, which from time to time carried reworked detritus from the shallower-water "cherty" facies southward into the deeper-water "slaty" facies.

Evidence also exists for a shallow-water origin for the "cherty" facies of the Ironwood. An algal horizon is present at the base of the lowest "cherty" member, the "cherty" facies has a granular nature, and chert fragment layers and terrigenous detritus (see Schmidt, 1980) are present.

That the iron-formations are gradational into the thick, overlying graywacke-slate units is shown by iron-formation lithologies well up in the graywacke-slate on the Gogebic range (the Tyler Formation—see Schmidt, 1980), on the Cuyuna range and the Emily district (the Rabbit Lake Formation—see Marsden, 1972), and the Mesabi range (the Virginia Formation—see Morey, 1972b; 1973; Lucente and Morey, in press). Such vertical transitions suggest that some lateral intertonguing is also present in the subsurface, and that classic facies relationships might be valid here as well as lower in the Animikie column. Minor graywackes and mudstones may have been accumulating by turbidity current and pelagic sedimentation on the slope and beyond at the same time that the shelf and littoral sediments were being deposited. By-passing of the littoral and shelf terrigenous and chemical sediments by the sands

and muds of the graywackes and mudstones would have been necessary. The graywacke-slate formations have been interpreted as deeper-water turbidite sequences (Morey, 1967, 1973; Morey and Ojakangas, 1970; and Alwin, 1979), which were the products of increased regional tectonism. Any lateral equivalency with the shelf deposits would have ended early in the tectonic cycle when the major portion of the shelf underwent rapid subsidence, perhaps as a result of faulting, and was buried beneath a thick pile of graywacke, siltstone, and mudstone derived from the adjacent, rising (upfaulted?) landmasses. The sands in the graywackes, although texturally immature, consist of 50 to 75 percent quartz (Morey, 1973; Alwin, 1979). The graywacke-mudstone sequence becomes sandier upward; this coarsening trend is evident in the Rove, Virginia, and Tyler Formations and may be sedimentological evidence for the onset of the Penokean orogeny in this portion of the Lake Superior region. Sedimentation ended as the Animikie basin was elevated during that orogeny.

SUMMARY AND CONCLUSIONS

The sedimentational-tectonic history of the early Proterozoic in Wisconsin and Michigan began with a marine transgression onto the peneplaned Archean plutonic-volcanic basement and deposition of the Sunday Quartzite and the Bad River Dolomite in a tidally-influenced pre-Animikie depocenter (Larue, 1981b). A broad, regional uplift and tilting resulted in local erosion of one or both units along most of the Gogebic range.

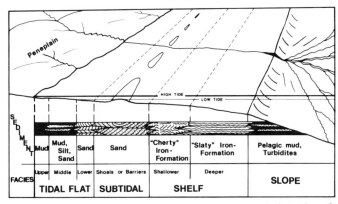

Figure 23. Sedimentational model showing lateral relationships of the siliciclastic tidal facies, the iron-formation facies on the shelf, and the deeper-water turbidite-mud facies. Thicknesses of units not drawn to scale.

This uplift was followed by subsidence, and a similar marine transgression into the new Animikie basin extended further north to the vicinities of the Mesabi and Gunflint ranges. The basin was rimmed by a broad (probably at least several kilometers wide) and shallow, but slowly subsiding shelf. On this shelf in areas that are now the Gogebic and Mesabi ranges, and probably also in the areas now occupied by the Cuyuna and Gunflint ranges, different tidal and shallow marine facies were being deposited side-by-side (Fig. 23). These were stacked vertically as the sea transgressed further over the adjacent, low-lying landmasses. Clastic terrigenous sediments comprised the tidal and shallow marine facies, and iron-formation formed further seaward, away from the land-derived clastics. Two major iron facies formed: a shallower-water, coarser-grained, clastic, reworked, and cross-bedded oxide-silicate "cherty" facies and in adjacent but somewhat deeper water, a finer-grained silicate-carbonate "slaty" facies. At and beyond the shelf edge, muds and graywacke sands accumulated, perhaps in small part contemporaneously with the shelf sediments; but then as the basin subsided rapidly and the adjacent landmasses were uplifted to provide great volumes of texturally and mineralogically immature detritus (Alwin, 1979), this deeper-water facies was deposited as a thick wedge upon the shelf deposits as well. The facies and formations as now stacked indicate, from the bottom to the top of the column, deposition in successfully deeper marine environments.

An alternative model can be designed in which the sandstone facies are high-energy beach or nearshore sands; the shale-siltstone-sandstone facies and the shale facies are tidally-influenced offshore deposits or are the result of energy alternations due to storms and fair-weather conditions (Raaf et al., 1977); and the iron-formations are lagoonal. However, such a model has several major problems:

(1) The vertical relations of these facies would necessitate a marine regression rather than a transgression, al-

though the basal units resting on older Precambrian rocks are apparently transgressive.

(2) The turbidite sequences above the iron-formations would necessitate deeper-water deposits laid down with a gradational contact upon lagoonal deposits, or else the graywacke-slate turbidite sequences must be reinterpreted as having formed in very shallow water.

(3) The two facies of iron-formation would both have to be lagoonal in origin in spite of their great lithologic differences.

(4) Classical Waltherian facies relationships for a stable tectonic regime must be contramanded by a unique set of circumstances, or stratigraphic breaks must be postulated *ad hoc* without tangible evidence for their presence.

The tidal model suggested here, and the resultant facies interpretations which allow for a determination of the sites of iron-formation deposition, may be of significance elsewhere in the Lake Superior region, especially in the Gunflint and Cuyuna ranges. The model is wholly new and is unique in that it is the sedimentary units associated with the iron-formations, rather than the iron-formations themselves, that provide the evidence for positioning of the iron-formation relative to the shoreline of the Animikie Sea. In view of the bewildering array of hypotheses for the origin of banded iron-formations, such an approach seems fruitful.

However, the model essentially superimposes the tidal aspect onto White's (1954) facies model for the shallow-water origin of the rocks of the Mesabi range. The model also lends support to portions of Larue's (1981a) transgressive depositional model in which iron-formation on the Gogebic range was deposited seaward from the sandstone and argillite, but differs markedly in interpretation of the positioning of the land-derived sedimentary facies relative to the chemical iron-formation. It further substantiates Drever's (1974) geochemical model of the deposition of iron-formation on continental margins near upwelling silica- and iron-rich bottom waters. This model is in general agreement with Dimroth's (1979) model based on the Sokoman Iron Formation of Labrador; tidal currents and cross-bedding occur in the "cherty" facies whose site of deposition was interpreted as a shelf marginal bank. For thin, interbedded sequences of "micrite-type" iron formation (the "slaty" facies) and "cherty facies," Chauvel and Dimroth (1973) in an excellent, detailed study suggested accumulation on a broad, lagoonal platform landward of oolite banks.

The chemical aspects of iron mineral deposition, while obviously important to the origin of iron-formation, are not critical to this model. It does not matter whether the source of the iron was volcanism or weathering. It does not matter if the iron minerals were of biogenic origin (e.g., La Berge, 1973; Schmidt, 1980) or not. Similarly, it is not of prime importance whether the iron minerals were precipitated as iron minerals, or were the result of the replacement by iron and silica during diagenesis of original carbonates as has been proposed for various banded iron-formations by Gross (1972), Dimroth (1979), Kimberley (1979), Markum and Randazzo (1980), and others. The model does add some physical sedimentational support to Holland's (1973) and Drever's (1974) models for upwelling of silica- and iron-rich waters at the shelf edge.

A tectonic framework involving rifting and block-faulting along the Great Lakes Tectonic Zone seems likely in the Lake Superior region during early Proterozoic time. Rifting is further supported by the presence of volcanics above the iron-formation in the Cuyuna range (Schmidt, 1963, p. 33), by tuffs in the Gunflint Iron Formation (Goodwin, 1956; Morey, 1973), by the intertonguing of the Ironwood Iron Formation with volcanics to the east (Trent, 1973; Prinz, 1980), by volcanics in the pre-Animikie Denham Formation (Morey, 1978), and by tuffs in the Biwabik and in the overlying Virginia Formation (Lucente and Morey, in press). Van Schmus (1976) suggested that deposition of the entire Animikie sequence occurred in a back arc basin, with the magmatic arc to the south in Wisconsin, but because this basin apparently does not have an oceanic crustal basement, it could be interpreted as a foreland basin. Larue (1981a, 1981b) has suggested that this platformal shelf sedimentation and the basin sedimentation of the Marquette area to the east occurred on a rifted passive continental margin. Young (1982) envisioned deposition in a fault-bounded ensialic trough or aulacogen. The information gleaned from this investigation is not by nature the type that will help resolve the tectonic setting.

ACKNOWLEDGMENTS

This study was accomplished in part while on WAE status with the U.S. Geological Survey. Paul Daniels assisted with some fieldwork. Jean Olson and Elizabeth Palmer did drafting and paleocurrent plotting. Typing was done by Avis Hedin and Mary Ankarlo. Ken Moran reproduced the illustrations. Ralph Marsden contributed to the project through numerous discussions and by reading the manuscript. Wayne Plummer kindly led a fieldtrip into Minntac's taconite pits. G. B. Morey, Erich Dimroth, Robert Dott, Jr., and Dave Larue read the manuscript and made several valuable suggestions.

REFERENCES CITED

Aldrich, H. R., 1929, The geology of the Gogebic Iron Range of Wisconsin: Wisconsin Geological and Natural History Survey Bulletin, 71, 279 p.

Alwin, B. W., 1979, The sedimentology of the middle Precambrian Tyler Formation, northern Wisconsin and Michigan [abs.]: Geological Society of America Abstracts with Programs, p. 225.

Anderton, Roger, 1976, Tidal-shelf sedimentation: an example from the Scottish Dalradian: Sedimentology, v. 23, p. 429–458.

Balazs, R. J., and Klein, G. deV., 1972, Roundness-mineralogical relations of some intertidal sands: Journal of Sedimentary Petrology, v. 42, p. 425–433.

Baragar, W.R.A., and Scoates, R.F.J., 1981, The circum-Superior belt: a Proterozoic plate margin? in A. Kroner (ed.), Precambrian Plate Tectonics, Elsevier Publishing Company, Amsterdam, p. 297–330.

Bayley, R. W., and James, H. L., 1973, Precambrian iron formations of the United States: Economic Geology, v. 68, p. 934–959.

Beukes, N. J., 1977, Transition from siliciclastic to carbonate sedimentation near the base of the Transvaal Supergroup, northern Cape Province, South Africa: Sedimentary Geology, v. 18, p. 201–222.

Blatt, Harvey, Middleton, Gerard, and Murray, Raymond, 1980, Origin of Sedimentary Rocks: Englewood Cliffs, New Jersey, Prentice-Hall Inc., 782 p.

Brunn, V. von, and Mason, T. R., 1977, Siliciclastic carbonate tidal deposits from the 3000 m.y. Pongola Supergroup, South Africa: Sedimentary Geology, v. 18, p. 245–256.

Button, A., 1976, Iron formation as an end member in carbonate sedimentary cycles in the Transvaal Supergroup, South Africa: Economic Geology, v. 71, p. 193–201.

Cannon, W. F., and Gair, J. E., 1970, A revision of stratigraphic nomenclature of Middle Precambrian rocks in northern Michigan: Geological Society of America Bulletin, v. 81, p. 2843–2846.

Chauvel, J. J., and Dimroth, E., 1974, Facies types and depositional environment of the Sokoman Iron Formation, Central Labrador Trough, Quebec, Canada: Journal of Sedimentary Petrology, v. 44, p. 299–327.

Cloud, Preston, and Licari, G. R., 1972, Ultrastructure and geologic relations of some two-aeon old Nostocaccean algae from northeastern Minnesota: American Journal of Science, v. 272, p. 138–149.

DeJong, J. D., 1977, Dutch tidal flats: Sedimentary Geology, v. 18, p. 13–23.

Dimroth, Erich, 1979, Models of physical sedimentation of iron formations: in R. G. Walker (ed.), Facies Models, Geoscience Canada Reprint Series 1, p. 175–182.

Dolence, J. D., 1961, The Pokegama Quartzite in the Mesabi Range [Unpublished M.S. thesis]: Twin Cities, University of Minnesota, 72 p.

Dott, R. J., and Batten, R. L., 1971, Evolution of the Earth: New York, McGraw-Hill, 649 p.

Drever, J. I., 1974, Geochemical model for the origin of Precambrian banded iron formations: Geological Society of America Bulletin, v. 85, p. 1099–1106.

Elliott, T., 1978, Clastic shorelines: in H. G. Reading (ed.), Sedimentary Environments and Facies: New York, Elsevier, p. 143–177.

Eriksson, K. A., 1977, Tidal deposits from the Archean Moodies Group, Barberton Mountain Land, South Africa: Sedimentary Geology, v. 18, p. 257–281.

Eugster, H. P., and Chou, I. M., 1973, The depositional environments of Precambrian banded iron formations: Economic Geology, v. 68, p. 1144–1168.

Evans, G., 1965, Intertidal flat sediments and their environments of deposition in The Wash: Quarterly Journal of the Geological Society of London, v. 121, p. 209–245.

Goldich, S. S., Nier, A. O., Baadsgaard, H., Hoffman, J. H., and Krueger, H. W., 1961, The Precambrian geology and geochronology of Minnesota: Minnesota Geological Survey Bulletin, v. 41, 193 p.

Goodwin, A. M., 1956, Facies relations in the Gunflint Iron Formation: Economic Geology, v. 51, p. 565–595.

Gross, G. A., 1965, Geology of iron deposits in Canada: general geology and evaluation of iron deposits: Geological Survey of Canada Economic Geology Report 22, v. 1, 181 p.

—— 1972, Primary features in cherty iron formations: Sedimentary Geology, v. 7, p. 241–261.

—— 1972, Primary features in cherty iron formations: Sedimentary Geology, v. 7, p. 241–261.

—— 1973, The depositional environment of principal types of Precambrian iron formations: *in* UNESCO, Genesis of Precambrian iron and manganese deposits, Proceeding of the Kiev Symposium, 1970, Earth Sciences, v. 9, p. 15–21.

—— 1980, A classification of iron formations based on depositional environments: Canadian Mineralogist, v. 18, p. 215–222.

Grout, F. F., and Broderick, T. M., 1919, The magnetite deposits of the eastern Mesabi range, Minnesota: Minnesota Geological Survey Bulletin, v. 17, 58 p.

Grout, F. F., Gruner, J. W., Schwartz, G. M., and Thiel, F. A., 1951, Precambrian stratigraphy of Minnesota: Geological Society of America Bulletin, v. 62, p. 1017–1078.

Gruner, J. W., 1922, Organic matter and the origin of the Biwabik iron-bearing formation of the Mesabi Range: Economic Geology, v. 17, p. 405–460.

—— 1923, Algae, believed to be Archean: Journal of Geology, v. 31, p. 146–148.

Holland, H. D., 1973, The oceans: A possible source of iron in iron formations: Economic Geology, v. 68, p. 1169–1172.

Hotchkiss, W. O., 1919, Geology of the Gogebic Range and its relation to recent mining developments: Engineering and Mining Journal, v. 108, p. 501–507.

Huber, N. K., 1959, Some aspects of the origin of the Ironwood Iron Formation of Michigan and Wisconsin: Economic Geology, p. 82–118.

Irving, R. D., and Van Hise, C. R., 1892, The Penokee iron-bearing series of Michigan and Wisconsin: U.S. Geological Survey Monograph, v. 19, 534 p.

James, H. L., 1954, Sedimentary facies of iron formation: Economic Geology, v. 49, p. 235–294.

Johnson, H. D., 1975, Tide- and wave-dominated inshore and shoreline sequences from the late Precambrian, Finnmark, North Norway: Sedimentology, v. 22, p. 45–74.

—— 1977, Shallow marine sand bar sequences: an example from the late Precambrian of North Norway: Sedimentology, v. 24, p. 245–270.

—— 1978, Shallow siliciclastic seas: *in* H. G. Reading (ed.), Sedimentary Environments, New York, Elsevier, p. 207–258.

Kimberly, M. M., 1979, Origin of oolitic iron formations: Journal of Sedimentary Petrology, v. 49, p. 111–132.

Klein, G. deV., 1963, Bay of Fundy intertidal zone sediments: Journal of Sedimentary Petrology, v. 33, p. 844–854.

—— 1968, Intertidal zone sedimentation, Minas Basin North Shore, Bay of Fundy, Nova Scotia: *in* A. E. Maroulies and R. C. Steers (eds.), National Symposium on ocean science and engineering of the Atlantic shelf, Marine Technological Society Transactions, p. 91–107.

—— 1970, Depositional and dispersal dynamics of intertidal sand bars: Journal of Sedimentary Petrology, v. 40, p. 1095–1127.

—— 1971a, A sedimentary model for determining paleotidal range: Geological Society of America Bulletin, v. 82, p. 2585–2592.

—— 1971b, Tidal origin of a Precambrian quartzite—The lower fine-grained quartzite (Middle Dalradian) of Islam, Scotland: Reply: Journal of Sedimentary Petrology, v. 41, p. 886–889.

—— 1972, Determination of paleotidal range in clastic sedimentary rocks: 24th International Geological Congress, Section 6, p. 397–405.

—— 1977a, Tidal circulation model for deposition of clastic sediment in epeiric and mioclinal shelf seas: Sedimentary Geology, v. 18, p. 1–12.

—— 1977b, Clastic Tidal Facies: Champaign, Illinois, Continuing Education Publication Company, 149 p.

Klein, G. deV., and Ryer, T. A., 1978, Tidal circulation patterns in Precambrian, Paleozoic, and Cretaceous epeiric and mioclinal shelf seas: Geological Society of America Bulletin, v. 89, p. 1050–1058.

Komar, P. D., 1976, Beach Processes and Sedimentation: New Jersey, Prentice-Hall, 429 p.

LaBerge, G. L., 1973, Possible biologic origin of Precambrian Iron Formations: Economic Geology, v. 68, p. 1098–1109.

Larue, D. K., 1981a, The early Proterozoic pre-iron formation Menominee Group siliciclastic sediments of the southern Lake Superior region: evidence for sedimentation in platform and basinal settings: Journal of Sedimentary Petrology, v. 51, p. 397–414.

—— 1981b, The Chocolay Group, Lake Superior region, USA: Sedimentologic evidence for deposition in basinal and platformal settings on an early Proterozoic craton: Geological Society of America Bulletin, v. 92, p. 417–435.

Larue, D. K., and Sloss, L. L., 1980, Early Proterozoic sedimentary basins of the Lake Superior region: Summary: Geological Society of America Bulletin, Part I, p. 450–452.

Leith, C. K., 1903, The Mesabi iron-bearing district of Minnesota: U.S. Geological Survey Monograph 43, 316 p.

Leith, C. K., Lund, R. J., and Leith, Andrew, 1935, Pre-Cambrian rocks of the Lake Superior region, a review of newly discovered geologic features, with a revised geologic map: U.S. Geological Survey Professional Paper 184, 34 p.

Lucente, M. E., and Morey, G. B., in press, Stratigraphy and sedimentology of the lower Proterozoic Virginia Formation (Animikie Group), northern Minnesota: Minnesota Geological Survey, Report Investigations: No. 28, 28 p.

Markum, C. D., and Randazzo, A. F., 1980, Sedimentary structures in the Gunflint Iron Formation, Schreiber Beach, Ontario: Precambrian Research, v. 12, p. 287–310.

Marsden, R. W., 1972, Cuyuna district: *in* P. K. Sims and G. B. Morey (eds.), Geology of Minnesota: A Centennial Volume, Minnesota Geological Survey, p. 227–239.

McCave, I. N., 1970, Deposition of fine-grained suspended sediment from tidal currents: Journal of Geophysical Research, v. 75, p. 4151–4159.

Mengel, J. G., 1973, Physical sedimentation in Precambrian cherty iron formations of the Lake Superior type: *in* G. S. Amstutz and A. J. Bernard (eds.), Ores in Sediments, Berlin, Springer-Verlag, p. 179–193.

Morey, G. B., 1967, Stratigraphy and sedimentology of the Middle Precambrian Rove Formation in northeastern Minnesota: Journal of Sedimentary Petrology, v. 37, p. 1154–1162.

—— 1972a, Middle Precambrian general geologic setting: *in* P. K. Sims and G. B. Morey (eds.), Geology of Minnesota: A Centennial Volume, Minnesota Geology Survey, p. 199–203.

—— 1972b, Mesabi Range: *in* P. K. Sims and G. B. Morey (eds.), Geology of Minnesota: A Centennial Volume, Minnesota Geological Survey, p. 204–217.

—— 1973, Stratigraphic framework of Middle Precambrian rocks in Minnesota: *in* G. M. Young (ed.), Huronian stratigraphy and sedimentation, Geological Association of Canada, Special Paper 12, p. 211–249.

—— 1978, Lower and Middle Precambrian stratigraphic nomenclature for east-central Minnesota: Minnesota Geological Survey, Report Investigations, v. 21, 52 p.

—— 1979, Stratigraphic and tectonic history of east-central Minnesota: *in* Field trip guidebook for stratigraphy, structure, and mineral resources of east-central Minnesota: Minnesota Geological Survey guidebook series No. 9, p. 13–28.

Morey, G. B., and Ojakangas, R. W., 1970, Sedimentology of the Middle Precambrian Thomson Formation, east-central Minnesota: Minnesota Geological Survey, Report Investigations, v. 13, 32 p.

Pettijohn, F. J., Potter, P. E., and Siever, Raymond, 1972, Sand and Sandstone: New York, Springer-Verlag, 618 p.

Prinz, W. C., 1967, Pre-Quaternary geologic and magnetic map and sections of part of the eastern Gogebic iron range, Michigan: U.S. Geological Survey Miscellaneous Geological Investigation Map I-497.

—— 1980, Geologic map of the Gogebic Range—Watersmeet area, Gogebic and Ontonagon Counties, Michigan: U.S. Geological Survey Miscellaneous Geological Investigation Map I-1365.

Raaf, J.F.M. de, and Boersma, J. R., 1971, Tidal deposits and their sedimentary structures: Geologie en Mijnbouw, v. 50, p. 479–504.

Raaf, J.F.M. de, Boersma, J. R., and Gelder, A. van, 1977, Wave-generated structures and sequences from a shallow marine succession, Lower Carboniferous, County Cork, Ireland: Sedimentology, v. 24, p. 451–483.

Reineck, H. E., 1972, Tidal flats: *in* J. K. Rigby and W. K. Hamblin (eds.), Recognition of ancient sedimentary environments, Society of Economic Paleontologists and Mineralogists Special Publication 16, p. 146–159.

Reineck, H. E., and Singh, I. B., 1972, Genesis of laminated sand and graded rhythmites in storm-sand layers of shelf mud: Sedimentology, v. 18, p. 123–128.

—— 1975, Depositional Sedimentary Environments: Berlin, Springer-Verlag, 439 p.

Reineck, H. E., and Wunderlich, Friedrich, 1968, Classification and origin of flaser and lenticular bedding: Sedimentology, v. 11, p. 99–104.

Sakamato, Takao, 1950, The origin of the Pre-Cambrian banded iron ores: American Journal of Science, v. 248, p. 449–474.

Schmidt, R. G., 1980, The Marquette Range Supergroup in the Gogebic Iron District, Michigan and Wisconsin: U.S. Geological Survey Bulletin 1460, 96 p.

Shegelski, R. J., 1980, Stratigraphy of the Gunflint Formation, Current River Area, Thunder Bay [abs.]: 26th Institute on Lake Superior Geology Proceedings and Abstracts, p. 28.

Simonson, B. M., 1982, Sedimentology of Precambrian iron formation with special reference to the Sokoman Formation and associated deposits of northeastern Canada [unpublished Ph.D. thesis]: Baltimore, The John Hopkins University, 359 p.

Sims, P. K., Card, K. D., Morey, G. B., and Peterman, Z. E., 1980, The Great Lakes tectonic zone—A major crustal structure in central North America: Geological Society of America Bulletin, Part I, v. 91, p. 690–698.

Swett, Keene, Klein, G. DeV., and Smit, D. E., 1971, A Cambrian tidal sand body—the Eriboll Sandstone of Northwest Scotland: An ancient-recent analog: Journal of Geology, v. 79, p. 400–415.

Tankard, A. J., and Hobday, D. K., 1977, Tide-dominated back-barrier sedimentation, early Ordovician Cape Basin, Cape Peninsula, South Africa: Sedimentary Geology, v. 18, p. 135–160.

Thompson, R. W., 1968, Tidal flat sedimentation on the Colorado River delta, northwestern Gulf of California: Geological Society of America Memoir 107, 133 p.

Trent, V. A., 1973, Geologic map of the Marenisco and Wakefield NE Quadrangles, Gogebic County, Michigan: U.S. Geological Survey open-file map.

Tyler, S. A., and Twenhofel, W. H., 1952, Sedimentation and stratigraphy of the Huronian of Upper Michigan, Parts I and II: American Journal of Science, v. 258, p. 1–27 and p. 118–151.

Van Hise, C. R., 1901, The iron ore deposits of the Lake Superior region: U.S. Geological Survey Annual Report 21, Part 3, p. 305–434.

Van Hise, C. R., and Leith, C. K., 1911, The geology of the Lake Superior region: U.S. Geological Survey Monograph 52, 641 p.

Van Schmus, W. R., 1976, Early and middle Proterozoic history of the Great Lakes area, North America: Philosophical Transactions of the Royal Society of London, Series A, v. 280, p. 605–628.

Van Straaten, L.M.J.U., 1959, Minor structures of some Recent littoral and neritic sediments: Geologie en Mijnbouw, v. 21, p. 197–211.

Watchorn, M. B., 1980, Fluvial and tidal sedimentation in the 3000 Ma Mozaan Basin, South Africa: Precambrian Research, v. 13, p. 27–42.

White, D. A., 1954, The stratigraphy and structure of the Mesabi Range, Minnesota: Minnesota Geological Survey Bulletin 38, 92 p.

Winchell, N. H., 1893, Twentieth annual report for the year 1891: Minnesota Geological Natural History Survey, 344 p.

Wolff, J. F., 1917, Recent geologic developments on the Mesabi iron range, Minnesota: American Institute of Mining and Metallurgical Engineering Transactions, v. 56, p. 142–169.

Yeo, R. K., and Risk, M. J., 1981, The sedimentology, stratigraphy and preservation of intertidal deposits in the Minas Basin System, Bay of Fundy: Journal of Sedimentary Petrology, v. 51, p. 245–260.

MANUSCRIPT ACCEPTED BY THE SOCIETY MARCH 4, 1983

Geological Society of America
Memoir 160
1983

Lower Proterozoic volcanic rocks and their setting in the southern Lake Superior district

Jeffrey K. Greenberg
Bruce A. Brown
Wisconsin Geological and Natural History Survey
1815 University Avenue
Madison, Wisconsin 53706

ABSTRACT

Studies of lower Proterozoic volcanic rocks in Wisconsin and northern Michigan reveal the existence of two different Penokean-age (about 1,900 to 1,800 m.y. old) geologic terranes. The terranes are in contact along the east-west trending Niagara fault and are confined between Archean craton to the north and progressively younger Proterozoic magmatic provinces to the south. The northern terrane shows some similarity to Andean-type continental margins. The geologic history of the area includes evidence of rifting, continental arc volcanism, and later orogenesis (collision?). Rocks in this environment were sutured to a southern terrane, the Penokean volcanic belt, which developed from an island arc and basin-type environment, probably flanking the continental margin.

The northern Penokean terrane contains thick units of sedimentary rocks, both platformal sequences and turbidites. Volcanic units are less abundant than sedimentary rocks and are typified by basalt flows and by lesser amounts of basaltic and rhyolitic volcaniclastic rocks. Penokean andesites are almost entirely absent from the northern terrane. Penokean calc-alkalic plutonic rocks are rare, whereas gabbroic sills and mafic layered complexes are common. The volcanic rocks display bimodal, tholeiitic trends with high iron enrichment. During what is considered to be the peak of the Penokean orogeny, the northern terrane rocks were affected by multiple deformations and metamorphic episodes. Tectonic styles in this region result from Archean basement uplifts and gneiss domes that exerted vertical stresses. Gneiss domes are outlined in the surrounding rocks by nodal patterns of metamorphic mineral zones.

Recent work and the integration of previous studies in the Penokean volcanic belt have shown the contrast between this region and the northern terrane. South of the Niagara fault, metavolcanic rocks are much more abundant than metasedimentary rocks. Sedimentary units appear to have been derived from varied, discontinuous sources and included conglomerates, graywackes, argillites, graphitic shales, tuffaceous sandstones, and dolomites. Penokean calc-alkalic volcanic suites and calc-alkalic plutons are uniquely abundant in the Penokean volcanic belt. However, minor quantities of tholeiitic volcanic and intrusive rocks occur within calc-alkalic suites in northeastern Wisconsin, just south of the Niagara fault. This area may have been the site of interarc magma genesis.

Tectonism in the Penokean volcanic belt can be distinguished by the lack of Archean basement, mantled gneiss domes, and metamorphic nodes. Greenschist-facies metamorphism was widespread throughout the belt. Higher-grade metamorphism was restricted to the vicinity of plutons and areas of locally intense deformation.

In spite of similarities, some features of modern plate-tectonic orogenies are not displayed by the Penokean. This leads to the conclusion that the Penokean may represent a transitional style of tectonism, not like Archean, nor exactly like modern plate-tectonic activity.

INTRODUCTION

Lower Proterozoic volcanic and associated rocks constitute a very large part of the exposed Precambrian bedrock in northern Wisconsin. These rocks can be viewed as making up a major tectonic belt, perhaps similar to granite-greenstone belts in Archean shields. Simplistic conclusions as to the character of this Penokean-age (approximately 1,860 to 1,800 m.y. old) belt are unrealistic, however, as large amounts of geologic data from many areas within the belt have never been collectively described and interpreted. This is due in part to the relative inaccessibility of much of the basic data in unpublished theses. We have integrated this information in order to compare Penokean rock chemistry and tectonic setting with more modern, Archean, and other Proterozoic volcanic suites. The result of our study is the derivation of a tectonic-

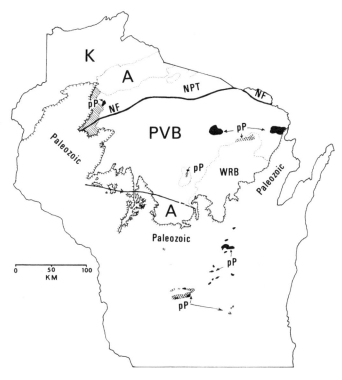

Figure 1. Major tectonic terranes in Wisconsin. Map modified from the page-size bedrock geology map of Wisconsin (Wisconsin Geological and Natural History Survey, 1981). K, Keweenawan terrane; A, Archean terranes; NPT, northern Penokean terrane; PVB, Penokean volcanic belt (southern terrane); NF, Niagara fault; WRB, Wolf River batholith; pP, post-Penokean; 1,760-m.y.-old magmatic rocks in black; quartzites cross-hatched.

petrologic framework for Penokean volcanism that best fits the available data.

The nature of tectonism during Proterozoic time places a major constraint on possible models for the origin of Penokean volcanic rocks. Some investigators have proposed that modern plate-tectonic processes have operated from Archean time (Burke and others, 1976; Condie, 1978). However, modern tectonic regimes often appear significantly different from those of the Archean, and any constant process of crustal evolution must have involved changes in structure and crustal composition. A thin, relatively mafic crust is often invoked for the Archean in contrast to thicker, more differentiated crust for the more recent Earth. Muehlburger (1980) noted that this thin crust and greater heat flow may have resulted in more rapid plate interactions in the early Precambrian than today. Without some unknown catastrophic change in dynamic processes, it follows that Proterozoic orogenies may in some ways be distinct from Archean tectonics and modern plate tectonics. The geology and chemical characteristics of lower Proterozoic volcanic rocks in Wisconsin may therefore provide a clearer understanding of the Penokean "Orogeny" (Goldich and others, 1961) and how it compares with modern and Archean tectonism.

PRECAMBRIAN GEOLOGIC SETTING

The Precambrian geology of northern Wisconsin and surrounding areas is characterized by several tectonic terranes, each consisting of a distinct group of rock units. These terranes are outlined on Figure 1.

Keweenawan-age (about 1,000 m.y. old) volcanic, sedimentary, and minor mafic intrusive rocks are the products of continental rifting in the extreme northwestern part of Wisconsin. A terrane including lower Proterozoic units and Archean-age granites, gneisses, and metavolcanic rocks (Greathead, 1975) lies southeast of the Keweenawan units. In the southern part of Michigan's Upper Peninsula and in northwestern Wisconsin, Archean gneisses and granites were reworked into domes during deformation (about 1,800 m.y. ago) with the overlying lower Proterozoic sedimentary and volcanic units (Sims, 1980). This terrane (referred to here as the northern Penokean terrane, NPT) is in fault-contact on the south with the dominantly volcanic and plutonic units of the Penokean volcanic belt (PVB).

Van Schmus (1976) recognized the distinctions between the two Penokean terranes and informally referred to the volcanic belt as a "volcano-plutonic belt." Larue and Sloss (1980) preferred the term "magmatic terrane" for the volcanic belt. All these designations emphasize that major 1,850-m.y.-old magmatic activity was widespread throughout the southern terrane in contrast with the predominantly sedimentary character of the northern Penokean terrane. In addition to the differences in rock types, the two terranes also contrast in age, the distribution of metamorphic facies, and structural style. The Niagara fault is the structure dividing the northern Penokean terrane from the Penokean volcanic belt. Deformational effects and metamorphism commonly intensify on either side toward the fault, but the most intense effects are not always restricted to the fault zone (Dutton, 1971).

Rocks of the Penokean volcanic belt are intruded on the southeast by alkalic granitic rocks of the Wolf River batholith (1,500 m.y. old). Along the southern margin of the volcanic belt, scattered exposures of Archean gneiss are restricted to south of a tectonic boundary (fault?) defined by changes in structural grain and geophysical anomalies (Fig. 1). The area south of this boundary is a structural domain unlike the rest of the Penokean volcanic belt and thus may constitute a third terrane involving Penokean magmatism and deformation (Maass and others, 1980). The southernmost exposures of Precambrian crystalline rock in Wisconsin are part of a 1,760-m.y.-old terrane of rhyolite, granite, and younger quartzose sedimentary units (Smith, 1978).

DISTRIBUTION AND CHARACTER OF PENOKEAN VOLCANIC ROCKS

Source of Information

Characteristics (excluding chemical compositions) of various Penokean volcanic suites are summarized below for the different areas of investigation where data are available (Fig. 2). These areas and the data sources utilized include:

Northern Penokean Terrane. The Marquette Range Supergroup (Weir, 1967; Dutton, 1971; Dann, 1978; Cudzilo, 1978).

Penokean Volcanic Belt. Northeast Wisconsin–Quinnesec area (Froelich, 1953; Thompson, 1955; Fulweiler, 1957; Prinz, 1959; Cain, 1962; Hall, 1971; Davis, 1977; Cummings, 1978; Cudzilo, 1978).

Rhinelander-Monico-Crandon area (Schriver, 1973; Venditti, 1973; Bowden, 1978; Schmidt and others, 1978; J. Hallberg, unpub. data).

Mountain area (Mancuso, 1960; Lahr, 1972; Motten, 1972; Cudzilo, 1978).

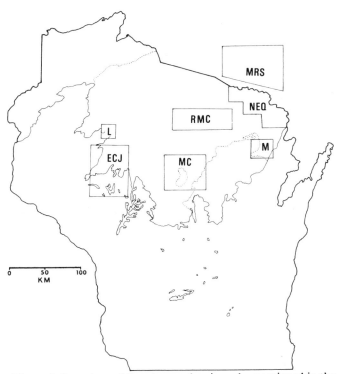

Figure 2. Locations of Penokean volcanic rocks mentioned in the text. MRS, Marquette Range Supergroup; NEQ, northeast Wisconsin–Quinnesec area; M, Mountain area; RMC, Rhinelander-Monico-Crandon area; MC, Marathon County; L, Ladysmith area; ECJ, Eau Claire-Chippewa-Jump Rivers area.

Marathon County (Voight, 1970; LaBerge and Myers, in prep.).

Ladysmith area (Edwards, 1920; May, 1977).

Eau Claire-Chippewa-Jump Rivers area (Myers and others, 1980).

In addition to these lower Proterozoic suites, geochemical data from Archean greenstones in northwestern Wisconsin (Greathead, 1975) were used for comparison with the younger rocks.

General stratigraphic relationships are known among the volcanic formations in the relatively well exposed Marquette Range Supergroup of the northern Penokean terrane (Table 1). Unfortunately, the typically limited bedrock exposure throughout the Penokean volcanic belt provides only a vague understanding of stratigraphy.

DESCRIPTION OF STUDY AREAS

Marquette Range Supergroup. Volcanic units exposed in the northern Penokean terrane include the Clarksburg, Hemlock, and Badwater Formations (Cudzilo, 1978; Dann, 1978) and the Emperor Complex (Dann, 1978). The Clarksburg and Hemlock Formations are approximate stratigraphic equivalents that underlie and are in part (Clarksburg Fm) interbedded with the Michigamme For-

TABLE 1.　STRATIGRAPHIC POSITION OF PENOKEAN VOLCANIC UNITS

Northern Penokean Terrane			Penokean Volcanic Belt		
West (Gogebic Range)		East (Iron, Dickenson Counties)	East	Central	West
(no equivalent)		Paint River Gp.	(no equivalent rocks)		
Marquette Range Supergroup — Baraga Gp.	Tyler Slate	*Michigamme Formation:* Badwater Greenstone Michigamme Slate Fence River Fm. Hemlock/Clarksburg (1900 m.y.)	Quinnesec Formation (1850 m.y.)	Marathon County Volcanics (1850 m.y.)	Eau Claire River Volcanics (1850 m.y.)
		Goodrich Quartzite			
Menominee Gp.	Ironwood Iron Fm. Emperor Volcanics	Vulcan Iron Fm.			
	Palms Quartzite	Felch Formation			
Chocolay Gp.	Bad River Dolomite	Randville Dolomite	(no equivalent rocks)		
	Sunday Quartzite	Sturgeon Quartzite			
		Fern Creek Fm.			

Note: Volcanic Units underlined, modified from Sims, 1976.

mation (Bayley and others, 1966). The Badwater Formation is stratigraphically above the Michigamme Formation. All three volcanic units are members of the Baraga Group. The Emperor Complex is probably stratigraphically lower than or roughly equivalent to the Clarksburg and Hemlock Formations. This uncertainty in position is due to problems in regional stratigraphic correlation among depositional basins (Larue and Sloss, 1980). Emperor Complex volcanic units underlie and are interbedded with Ironwood Iron Formation which is below the Tyler Slate, an equivalent of the Michigamme Slate. The Emperor Complex includes both a series of tholeiitic sills intruding a basalt sequence and also a sequence of calc-alkalic volcanics (the Wolf

Mountain Creek Formation). According to Dann (1978), the Wolf Mountain Creek Formation grades upward from a clastic sedimentary base through volcaniclastic sediments to interlayered massive and pillowed flows. Some hyaloclastites and flow breccias are also present in the upper units. Compositionally, the volcanic rocks range from basalt and andesite to dacite.

The Clarksburg, Hemlock, and Badwater Formations are comprised of similar volcanic units, though the Clarksburg contains no felsic members (Dann, 1978). It consists of mafic flows, tuffs, and coarse fragmental rocks that overlie iron formation. The Hemlock Formation consists of 600 m of basaltic flows, siliceous tuffs, and rhyolites. Badwater units are the youngest volcanic rocks, but also consist of pillowed mafic flows, mafic and felsic tuffs, and coarse fragmental rocks.

Metamorphism has destroyed most original textures in the Clarksburg, Hemlock, and Badwater Formations. Typical mineral assemblages include plagioclase, chlorite, epidote, calcite, hornblende, and in some places garnet, indicative of more intense metamorphism. However, the Emperor Complex, particularly the Wolf Mountain Creek Formation, displays many primary textural features and lower greenschist-facies metamorphic assemblages (Dann, 1978).

Northeast Wisconsin–Quinnesec Volcanics.

The best exposures in the Penokean volcanic belt occur in eastern Florence and northern Marinette Counties (NEQ, Fig. 2). Most of the metavolcanic materials in this area have been grouped in the Quinnesec Formation. This formation consists mainly of basaltic flow rocks interlayered with andesites, rhyolites, pyroclastics, epiclastics, and minor iron formations in the southern part of the area (Cummings, 1978). To the north, near the Niagara fault, the Quinnesec contains massive and pillow basalts with interlayered rhyolite flows and tuffs. In this same area, there are many bodies of gabbro and ultramafic rocks (Hall, 1971). These units are more intensely deformed and metamorphosed than rocks of the Marquette Range Supergroup which occur north of the Niagara fault (Bayley and others, 1966; Dutton, 1971). Jenkins (1973) recognized four volcanic units, including the Quinnesec, separated by faults in the eastern part of the area. His other three units are the Beecher Formation, typically porphyritic flows with sodic-plagioclase phenocrysts; the Pemene Formation, consisting of spherulitic sodic-rhyolite and rhyodacite; and the McAlister Formation, which consists largely of mafic coarse-fragmental rocks. Sedimentary rocks are generally scarce but occur as thin interflow units throughout the area. Sedimentary rock types include iron formation, graphitic mudstone, impure marble, sandy dolomite, quartzite, and metagraywacke (Dutton, 1971; Davis, 1977; Cummings, 1978).

Much of the area of Quinnesec and related deposition was invaded by large volumes of plutonic rocks during the Penokean orogeny. As a result, many exposures of volcanic and sedimentary rocks show amphibolite-facies metamorphic effects (metamorphic hornblende, garnets, and so forth).

Rhinelander-Monico-Crandon Area.

Outcrops in the Rhinelander-Monico-Crandon area (RMC, Fig. 2) are restricted to scattered exposures, many of which are dioritic to granitic intrusive rocks. Data on the volcanic rocks are limited to the vicinity of Monico and to exploration drill cores. Volcanic rocks in the Monico area (Schriver, 1973; Venditti, 1973) consist of a sequence of massive and pillowed mafic flows which are porphyritic to amygdaloidal. Phenocrysts were originally plagioclase and, more rarely, pyroxene. These flow units have been metamorphosed to assemblages of actinolite, chlorite, and epidote, and appear to overlie a succession of mixed felsic to intermediate tuffs, breccias, and mafic to intermediate flows. The breccias and tuffs have been tentatively correlated (M. Mudrey, personal commun.) with rocks exposed and drilled at the Pelican massive sulfide deposit, near Rhinelander and just west of Monico. In his study of the Pelican deposit, Bowden (1978) described foot-wall rocks as porphyritic and amygdaloidal mafic flows, flow breccias, tuffs, and epiclastic sediments. The breccias and tuffs are highly altered near the ore zone. Metamorphosed and hydrothermally altered mineral assemblages include the groundmass phases: iron-chlorite, epidote, zoisite-clinozoisite, actinolite, quartz, albite, calcite, and pyrite. Sausseritized plagioclase laths comprise most of the relict phenocrysts. In the hanging wall, units are distinguished by the presence of chert fragments and cherty tuffs as well as intermediate flows and crystal tuffs. Tuffaceous and massive intermediate volcanic rocks drilled during exploration of the Crandon massive sulfide deposit (Schmidt and others, 1978) are thought to be part of the same sequence exposed in the Monico area (M. Mudrey, personal commun.). Metamorphism in the Rhinelander-Monico-Crandon area is lower greenschist facies, but it is locally higher near contacts with plutonic rocks.

Mountain Area.

The Waupee volcanic sequence or "formation" (Mancuso, 1960; Lahr, 1972) is exposed in the vicinity of Mountain, Oconto County, Wisconsin (M, Fig. 2). Most exposures show a trend near N. 55° E. with steep dips. Lahr (1972) divided the Waupee into three units. The basal unit consists of about 5,000 m of massive to porphyritic mafic flows. Typical mineral assemblages are dominated by hornblende and plagioclase. In some places, minor siliceous zones occur in tuffs and sediments interlayered with mafic flows. The middle member of the Waupee consists of quartzofeldspathic sediments with minor interlayered mafic flows. Lahr (1972) mentioned no thicknesses for the middle or upper units. The upper unit is made up of fine-grained and thinly laminated tuffs, with stratigraphic tops facing northwest. Locally, the Waupee rocks are intensely deformed and metamorphosed to middle amphibolite facies. These effects probably resulted from the close proximity to the Wolf River batholith, whose intrusion obscured much of the original nature of the Waupee volcanics.

Marathon County.

Relatively good exposures of Penokean volcanic rocks are found in Marathon County (MC, Fig. 2) (Voight, 1970; LaBerge and Myers, in prep.) Voight (1970) described the volcanic sequence of northwestern Marathon County as interlayered massive basalts and andesites, with no evidence of pyroclastic material. The basalts exhibit ophitic and amygdaloidal textures and are composed of epidote, chlorite, magnetite, and minor actinolite. According to Voight (1970), "andesites" commonly display trachytic textures with subhedral microlites of plagioclase and anhedral albite in a groundmass of epidote, chlorite, and opaques. Porphyritic andesites contain twinned and zoned sodic-plagioclase phenocrysts in a groundmass of subhedral albite, epidote, sericite, and opaques. In other parts of Marathon County, particularly central and eastern, rhyolites and felsic tuffs are exposed along with graywackes and minor iron formation (LaBerge and Myers, in prep.). Rhyolites are massive, often porphyritic, and occasionally display original flow banding. Coarse fragmental units and crystal and lapilli tuffs are also present in some areas.

Ladysmith Area. Much of the stratigraphic information from the western part of the volcanic belt comes from drill core data of the Ladysmith massive sulfide deposit (L, Fig. 2). The deposit is within a sequence of steeply dipping schistose volcanic and sedimentary rocks, truncated by granite intrusions (May, 1977). Rock types identified, mostly from drill core, include dacitic to rhyolitic crystal tuffs, massive dacitic to andesitic flows (now metamorphosed to actinolite schists), and rhyolitic tuffs, represented by quartz-sericite (±andalusite) schists.

Eau Claire-Chippewa-Jump Rivers Area. The southwesternmost exposures of the Penokean volcanic belt occur in the Eau Claire, Chippewa, and Jump River Valleys (ECJ, Fig. 2). Much of the geology in this region is summarized in Myers and others (1980). Rocks in the area include basaltic, andesitic, and rhyolitic flows, tuffs, and coarse fragmental rocks interstratified with siliceous volcanogenic sediments, slates, quartzites, and conglomerates. Myers and others (1980) have suggested that the volcanic rocks of the Eau Claire area are younger than and overlie amphibolite units. Volcanic rocks of the Jump River valley are described by Cummings (Myers and others, 1980) as basaltic and andesitic flows that are interbedded with andesitic to rhyolitic fragmental rocks. The rhyolitic flows and coarse fragmental materials lead Cummings to propose the existence of a felsic eruptive center in the Jump River area. Although no thicknesses were given, the extent of these rocks exposed suggests that they are at least several hundreds of metres thick (Myers and others, 1980).

AGE DATA

The few radiometric ages available for Penokean volcanic rocks in northern Wisconsin and the Upper Peninsula of Michigan thus far indicate that volcanism was probably earlier in the Marquette Range Supergroup (NPT) than in the Penokean volcanic belt (Van Schmus and Bickford, 1981). Felsic units in the Hemlock Formation are about 1,900 m.y. old (U-Pb, Van Schmus, 1976). U-Pb zircon ages of volcanic suites from three different areas within the Penokean volcanic belt are all about 1,850 m.y. (Van Schmus, 1980). This is roughly the age of rhyolites from the

Quinnesec Formation (Banks and Rebello, 1969), from Marathon County, and from Eau Claire County (Fig. 2). The zircon ages correlate well with Pb-Pb model ages of near 1830 m.y. for volcanogenic sulfide deposits located in the central and western parts of the Penokean volcanic belt (Stacey and others, 1977). Anorogenic rhyolites south of the main volcanic belt are near 1,760 m.y. old and represent a change in tectonic environment after the Penokean orogeny (Van Schmus and Bickford, 1981).

INTERPRETATION OF GEOCHEMICAL DATA

Because the many sources used in this data compilation probably vary in quality, chemical analyses were chosen according to certain criteria. For example, odd analyses, such as rocks described as basalt but containing 25% Al_2O_3 or 30% SiO_2 were rejected. This type of situation may arise from misidentification or major analytical errors. In addition, samples with element abundances totaling less than 95% or more than 102% were not incorporated in the compilation. Averaged compositions of mafic, intermediate, and felsic volcanic rocks from both Penokean terranes are shown in Table 2.

Several variation diagrams were tested for data interpretation, but only a few were selected to best display the relationships among different volcanic suites. Unfortunately, trace elements were not analyzed in most studies

TABLE 2. AVERAGE COMPOSITION OF VOLCANIC ROCKS FROM BOTH PENOKEAN TERRANES

	Northern Penokean Terrane					
	Rhyolite (68–72% SiO_2)		Andesite (57–62% SiO_2)		Basalt (47–51% SiO_2)	
		SD		SD		SD
SiO_2	69.19	1.89	59.20	2.78	48.80	1.35
Al_2O_3	12.24	0.84	14.13	1.11	14.56	1.48
Fe_2O_3*	5.43	0.90	9.93	0.76	14.23	1.93
TiO_2	0.39	0.12	1.13	0.01	1.54	0.49
MgO	1.25	0.95	4.58	0.50	5.78	2.00
CaO	0.72	0.92	4.53	0.96	8.19	3.20
Na_2O	2.26	1.78	5.54	0.90	2.73	0.75
K_2O	4.66	2.64	1.35	1.06	0.72	0.44
N	3		3		21	

	Penokean Volcanic Belt					
	Rhyolite (68–72% SiO_2)		Andesite (57–62% SiO_2)		Basalt (47–51% SiO_2)	
		SD		SD		SD
SiO_2	69.86	1.70	59.24	1.76	49.15	1.17
Al_2O_3	14.03	1.74	15.16	2.02	15.35	2.14
Fe_2O_3*	3.95	2.32	8.71	2.16	12.65	1.91
TiO_2	0.33	0.30	0.59	0.34	0.71	0.32
MgO	0.90	1.12	4.08	1.68	5.30	1.54
CaO	1.95	1.65	4.39	1.56	10.41	2.25
Na_2O	4.07	1.36	3.92	1.43	2.36	2.04
K_2O	2.50	1.76	1.39	0.96	0.62	0.51
N	8		16		27	

Note: Elements are in weight percent.
  N is the number of samples.
  * Total iron computed as Fe_2O_3.
  SD is standard deviation.

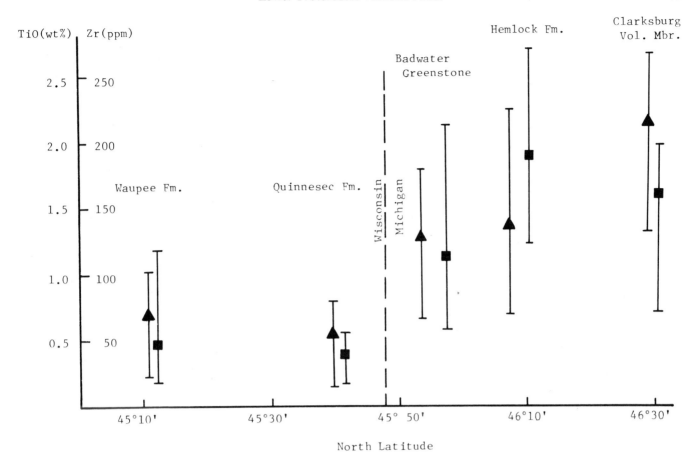

Figure 3. Plot of TiO$_2$ and Zr versus latitude from Cudzilo (1978). Solid triangles represent TiO$_2$ in weight percent. Solid squares represent Zr in parts per million. Vertical bars represent range of values for each unit.

and their usefulness as indicators of tectonic environments are thus limited. It was also not practical to interpret most plots of the more mobile elements (such as Al$_2$O$_3$, CaO, and the alkalis) which tend to be more affected by the influences of hydrous alteration. The plots eventually utilized for data comparison include: SiO$_2$ histogram, SiO$_2$ versus log$_{10}$K$_2$O/MgO, Ni versus MgO, total Fe versus MgO, Zr-Ti-Sr, and Ti versus Zr. Although it could be argued that alteration or other processes complicate geochemical interpretation, the present study was specifically intended to emphasize major similarities and differences among volcanic rock suites. In fact, a twofold empirical test validated the use of the data plots. On each diagram there is a good grouping of samples from the same area (and not always from the same study), and more important, there is general agreement among conclusions from the different plots.

Chemical differences between the two Penokean terranes are apparent on all the variation diagrams. Cudzilo (1978) illustrated this contrast with his Figure 10 (Fig. 3 this paper, by permission) of relative TiO$_2$ and Zr abundances. The contrast between volcanism in the two terranes is also the most striking feature of SiO$_2$ histograms (Fig. 4). In the

northern terrane, all Penokean volcanic units except the calc-alkalic portion of the Emperor Complex (Wolf Mountain Creek volcanics) are bimodal volcanic suites, devoid of andesites. Basaltic rocks (53% to 56% SiO$_2$ as an upper limit) predominate, but samples from all but the Clarksburg Formation include rhyodacite to rhyolite. Definite bimodality does not occur for any of the analyzed suites south of the Niagara fault. Unfortunately, the impression of bimodality can be given in some areas if sampling biases are not taken into consideration. An example of this relates to the Penokean volcanogenic sulfide deposits (May, 1977), where many samples were analyzed from near ore zones, in which case analyses of samples were biased toward the felsic side. Metamorphosed intermediate to felsic tuffs are host rocks in the Ladysmith (May, 1977), Crandon (Schmidt and others, 1978), and Pelican (Bowden, 1978) massive sulfide deposits. Although Van Schmus and Bickford (1981) have stated to the contrary, andesites are well represented in all areas of the Penokean volcanic belt and include rocks of widely variable metamorphic facies and deformational fabric. These intermediate volcanic rocks are defined by their SiO$_2$ contents (Fig. 4) as well as their overall chemical and physical character described in various

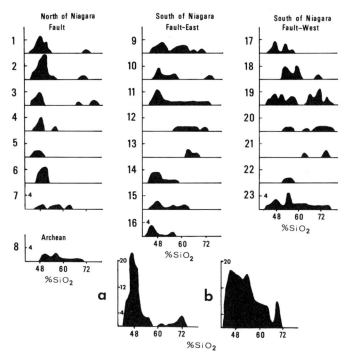

Figure 4. SiO₂ histograms for various Penokean, Wisconsin Archean, and composite volcanic suites. SiO₂ in weight percent. 1, Hemlock Formation (Cudzilo, 1978); 2, Hemlock Formation (Dann, 1978); 3, Badwater Formation (Cudzilo, 1978); 4, Badwater Formation (Dann, 1978); 5, Clarksburg Formation (Cudzilo, 1978); 6, Emperor Complex, tholeiitic (Dann, 1978); 7, Emperor Complex, calc-alkalic (Dann, 1978); 8, northwestern Wisconsin Archean volcanics (Greathead, 1975); 9, northeast Wisconsin–Quinnesec area (Davis, 1977); 10, northeast Wisconsin–Quinnesec area (Hall, 1971); 11, Quinnesec Formation (Cudzilo, 1978); 12, Beecher Formation (Cudzilo, 1978); 13, Pemene Formation (Cudzilo, 1978); 14, Mountain area, Waupee volcanics (Lahr, 1972); 15, Mountain area, Waupee volcanics (Cudzilo, 1978); 16, northeast Wisconsin–Quinnesec area (Cummings, 1978); 17, Monico area (Schriver, 1973); 18, Monico area (Venditti, 1973); 19, Rhinelander area, Pelican massive sulfide deposit (Bowden, 1978); 20, Marathon County (LaBerge and Myers, in prep.); 21, Ladysmith area, Ladysmith massive sulfide deposit (May, 1977); 22, Ladysmith area (Edwards, 1920); 23, Eau Claire-Chippewa-Jump Rivers area (Myers and others, 1980); a, composite of all samples north of the Niagara fault; b, composite of all samples south of the Niagara fault.

studies. The only Archean metavolcanic rocks studied in Wisconsin are north of the Penokean volcanic belt (Greathead, 1975). These Archean samples cover all silica ranges but are concentrated toward the basalt end.

In more modern environments, silica bimodality is usually equated with tensional tectonics (rifting) in continental or oceanic settings (Condie, 1976b). Silica distributions that include andesites usually imply arc volcanism, either continental or island. Modern analogues suggest that most magmas in the northern Penokean terrane originated in tensional environments, whereas those in the Penokean volcanic belt were orogenic products.

Regardless of some problems that might be caused by K₂O mobility, SiO₂ plotted against K₂O/MgO (Fig. 5) reveals some general relationships. Samples from modern calc-alkalic volcanic suites (continental and island arc) cluster close to or above a reference trend line defined by several volcanoes from the northwestern Philippine arc (DeBoer and others, 1980). Mafic suites, such as ocean-floor tholeiites, plot near the origin (Fig. 5a), whereas more differentiated suites trend toward higher K₂O values, but at slope angles consistent with their relative increase in K₂O. Anorogenic suites, such as continental flood-volcanics (Leeman and Vitaliano, 1976; Wright and others, 1973), and Keweenawan rift-volcanics (Green, 1972) show relative K₂O enrichment trends below the calc-alkalic line. Archean suites from various regions, including Wisconsin, typically follow trends of lower K₂O/MgO increase, above the calc-alkalic line. These trends reflect high MgO rather than low K₂O content.

With reference to K₂O content, none of the Penokean volcanic suites from Wisconsin and northern Michigan can really be considered as alkaline in any classification scheme. The northern Penokean terrane suites all conform to anorogenic, continental-tholeiite trends, although some of the felsic rocks from this region are more magnesian than those of modern analogues (Fig. 5a, 5b). This may be due to the specific nature of bimodal volcanism during the Penokean orogeny. The calc-alkalic Wolf Mountain Creek suite (Dann, 1978) forms a small field distinct from the rest, suggesting a magma genesis unique in this terrane.

South of the Niagara fault (Fig. 5c, 5d), differentiation trends are more complex. As Cudzilo (1978) observed, the Waupee volcanics are more alkaline than all other units examined in his study. At low SiO₂, the Waupee rocks parallel anorogenic suites. Samples from May (1977), Bowden (1978), and Cummings (1978) show the effects of alteration (magnesium metasomatism) during mineralization. All other suites from the central and western part of the volcanic belt including Monico (Schriver, 1973; Venditti, 1973), Marathon County (LaBerge and Myers, in prep.), and the Chippewa River valley (Myers and others, 1980) conform to calc-alkalic patterns and trends.

Further geochemical observations can be summarized as follows:

1. Volcanic rocks in the northern Penokean terrane plot in fields representing anorogenic tholeiites, usually similar to continental suites (Figs. 6, 7, 8). High iron enrichment is evident in Figure 6. Many of these samples overlap high-iron Archean greenstones but do not show an equivalent enrichment in MgO, characteristic of komatiites. In addition, Archean volcanic rocks are typified by higher Ni/MgO ratios (Fig. 7) than all but those Penokean suites containing sulfide mineralization. The Penokean volcanic rocks from both the northern and volcanic belt terranes are not appreciably different than some suites from

modern tectonic environments. However, with the prominent exception of SiO$_2$ contents, most chemical data suggest that volcanism in the Penokean volcanic belt, like that in the northern Penokean terrane, was somewhat distinct from volcanism during the Archean. Even for samples with equal SiO$_2$ contents, Penokean calc-alkalic suites contain generally lower amounts of total Fe, MgO, TiO$_2$ and Ni than Archean suites (Table 2; Condie, 1976a; Gunn, 1976).

2. A calc-alkalic character is evident for samples from throughout the Penokean volcanic belt shown on Figures 6, 7, and 8. The only major deviation from orogenic arc-type chemistry is exhibited by the Quinnesec and associated Beecher and Pemene Formations (Hall, 1971; Cudzilo, 1978). These samples are chemically similar to rocks from the Flin Flon and Hastings Proterozoic regions of Canada (Moore, 1977) which contain both calc-alkalic and tholeiitic units (Fig. 6). Quinnesec volcanism was conspicuously more complex than the rest of the Penokean volcanic belt. There is, in addition, the impression of possible iron enrichment in the northeast relative to the rest of the belt

(Fig. 6). This phenomenon probably indicates a fundamental change in the character of arc volcanism from southwest to northeast.

STRUCTURE OF THE VOLCANIC BELT

The structural complexities encountered in the northern Penokean terrane are discussed in Cannon (1973) and Klasner (1978), specifically for northern Michigan. The structural geology of the Penokean volcanic belt has never been studied in any detail. However, in spite of limited exposure, enough data, including much from the authors' own study, are now available to draw some general conclusions regarding the structural history of this area.

Although the typical style of deformation is not greatly different in the volcanic belt, the effects of large gneiss-dome development are only observed in the northern Penokean terrane, that is, north of the Niagara fault and its continuation to the west (Fig. 9). South of the fault, the most pervasive structural element is a penetrative foliation

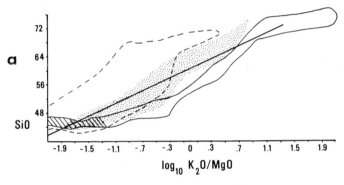

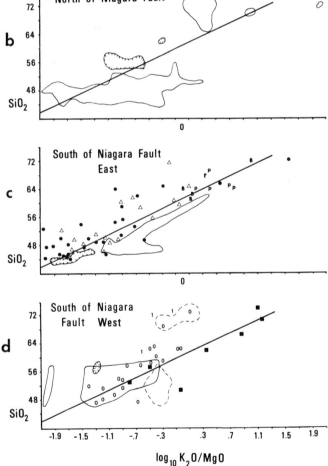

Figure 5. Plots of SiO$_2$ versus log$_{10}$K$_2$O/MgO. All values in weight percent. The line represents the trend of calc-alkalic suites from Philippine volcanoes (DeBoer and others, 1980). a. Solid-line field represents anorogenic volcanic suites including Snake River flood volcanics (Leeman and Vitaliano, 1976); Columbia River flood volcanics (Wright and others, 1973); Keweenawan volcanics (Green, 1972); and ocean-floor tholeiites, emphasized with diagonal ruling (Gunn, 1976). Stippled field represents calc-alkalic (orogenic) suites including Cascades arc (Turner and Verhoogen, 1960); Papuan arc (Heming, 1974); Marianas arc (Stern, 1979); Martinique (Gunn and others, 1974); Banda arc (Whitford and Jezek, 1979). Dashed-line field represents Archean volcanic suites including Vermilion, MN area (Schulz, 1980); Finland (Jahn and others, 1980); northwest Wisconsin (Greathead, 1975); and Abitibi belt (Jolly, 1977). b. Solid-line fields represent the Emperor tholeiitic, Clarksburg, Hemlock, and Badwater Formations. Hachured field represents the Emperor calc-alkalic suite. c. Solid-line field represents Waupee volcanics. Hachured field represents samples of Cummings (1978). Triangles represent Quinnesec Formation samples of Davis (1977). Solid dots represent Quinnesec Formation samples of Hall (1971) and Cudzilo (1978). Symbol B represents Beecher Formation samples. Symbol p represents Pemene Formation samples. d. Solid-line field represents Monico area samples. Hachured-line field represents Ladysmith area samples (Edwards, 1920). Dashed-line field represents Rhinelander area samples (Bowden, 1978). Solid squares represent

Marathon County Samples. Symbol 0 represents Eau Claire-Chippewa-Jump River area samples. Symbol 1 represents smith area samples (May, 1977).

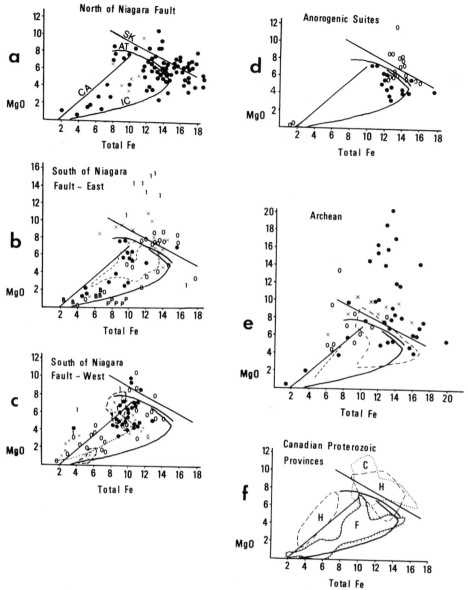

Figure 6. Plots of total iron versus MgO. All values in weight percent. Trends (from Moore, 1977) on all diagrams include SK, Skaergaard; AT, abyssal tholeiites; CA, cascades; and IC, Iceland. a. Solid dots represent the Emperor tholeiitic, Clarksburg, Hemlock, and Badwater volcanic Formations. Symbol x represents Emperor calc-alkalic suites. b. Dashed-line field represents Waupee samples. Solid dots represent Quinnesec Formation samples of Davis (1977). Symbol x represents Quinnesec Formation samples of Hall (1971). Symbol 0 represents Quinnesec Formation samples of Cudzilo (1978). Symbol B represents Beecher Formation samples. Symbol p represents Pemene Formation samples. Symbol 1 represents samples of Cummings (1978). c. Dotted-line represents trend of Papuan arc volcanics (Heming, 1974). Dashed-line field represents Marathon County samples. Solid dots represent Monico area samples. Symbol x represents samples of Rhinelander area samples (Bowden, 1978). Symbol 0 represents Eau Claire-Chippewa-Jump River area samples (Myers and others, 1980). Symbol 1 represents Ladysmith area samples (Edwards, 1920; May, 1977). d. Solid dots represent Columbia River flood-volcanics (Wright and others, 1973). Symbol 0 represents Snake River flood-volcanics (Leeman and Vitaliano, 1976). e. Dashed-line represents "primitive" calc-alkalic trend (Jolly, 1977). Dashed-line field represents Abitibi tholeiitic samples (Jolly, 1977). Solid dots represent Finnish greenstone belt samples (Jahn and others, 1980). Symbol x represents Vermilion greenstone belt samples (Schulz, 1980). Symbol 0 represents northwest Wisconsin samples (Greathead, 1975). f. All fields are of Canadian lower Proterozoic volcanic suites (Moore, 1977). H, dashed-line field represents Hastings area volcanics. C, dotted-line field represents Circum-Ungava belt volcanics. F, hachured-line field represents Flin Flon area volcanics.

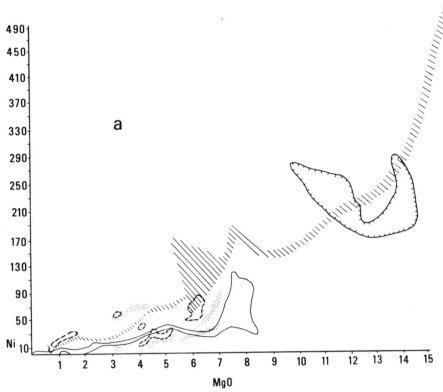

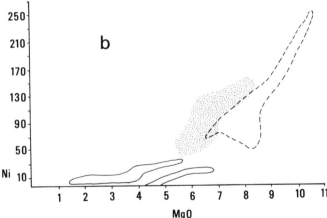

Figure 7. Plot of Ni versus MgO. Ni values in parts per million. MgO in weight percent. a. Solid-line field represents Quinnesec, Beecher, and Pemene Formations (Cudzilo, 1978). Dashed-line fields represent Clarksburg, Hemlock, and Badwater samples (Cudzilo, 1978). Stippled fields represent Waupee samples (Lahr, 1972; Cudzilo, 1978). Hachured-line field represents samples from Cummings (1978). Diagonally ruled field represents Archean suites (Condie, 1976a; Jahn and others, 1980; Schulz, 1980). b. Solid-line fields represent island arc suites (Heming, 1974; Gunn and others, 1974; Stern, 1979). Dashed-line field represents ocean-floor and mid-ocean ridge suites (Bougault and Hekinian, 1974). Stippled field represents continental flood volcanics (Leeman and Vitaliano, 1976).

associated with isoclinal folding during the Penokean orogeny. The folding is responsible for the steep to overturned dips of most bedded rocks in the belt. The axial trend of Penokean folding is east to northeast in the extreme western part of the belt, gently arching through the Marathon County–Rhinelander region, and east to southeast in Florence and Marinette Counties (Fig. 9). Major northeast- to east-trending faults are subparallel with the primary foliation and fragment the belt into many subregions or structural blocks. Fault displacement directions or magnitudes cannot be determined from the available data. Secondary features, such as crenulations, cross-cutting open folds, and folded lineations are common, particularly in areas of 1,850-m.y.-old plutonic activity. The secondary fold axial planes typically trend northwest to northeast.

Excellent examples of Penokean structures can be observed in northeastern Wisconsin. Here, anticlines and synclines several kilometres across have been mapped in both Penokean terranes. In the Marquette Range Supergroup (Bayley and others, 1966; Dutton, 1971), these large folds have been refolded and then faulted in a series of steps, exposing lower crustal levels to the south, up to the Niagara fault and the Quinnesec Formation. The Quinnesec Formation also contains large isoclinal folds, some of which are doubly plunging (Jenkins, 1973; Cummings, 1978). Cross-folds, well-developed lineations, flattened clasts, ductile shear zones, and other intense deformational features in the Quinnesec appear to be related to Penokean plutons rather than to any earlier complexity.

Due to the lack of continuous exposure, the regional

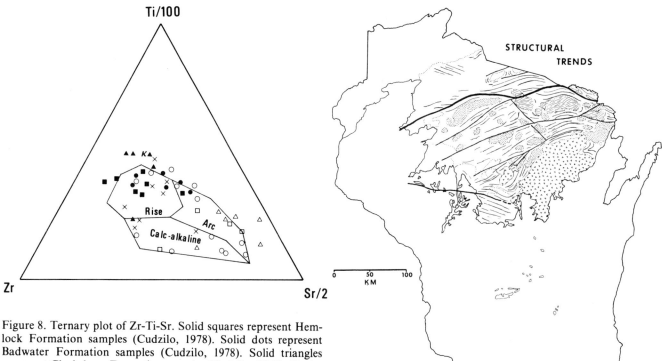

Figure 8. Ternary plot of Zr-Ti-Sr. Solid squares represent Hemlock Formation samples (Cudzilo, 1978). Solid dots represent Badwater Formation samples (Cudzilo, 1978). Solid triangles represent Clarksburg Formation samples (Cudzilo, 1978). Circles represent Quinnesec Formation samples (Cudzilo, 1978). Open triangles represent Waupee volcanics (Cudzilo, 1978). Open squares represent Monico area samples (J. Hallberg, unpub. data). Symbol *K* represents Keweenawan basalt average (Green, 1972). Symbol x represents Archean samples (Condie, 1976a; Cudzilo, 1978).

Figure 9. Structural trends in the Penokean terranes of northern Wisconsin. Heavy lines indicate major Penokean tectonic boundaries. Lighter lines indicate trends of faults and foliation. Penokean intrusive rocks are shown in pattern. Structural data and lithologic boundaries from Greenberg and Brown (1980) and Wisconsin Geological and Natural History Survey (1981).

structure of most of the western portion of the volcanic belt is poorly understood. Volcanogenic rocks in the Monico area have been refoliated near major east-west faults. Some of these rocks also show strong flattening and lineation near discrete plutons (Venditti, 1973). Bowden (1978) proposed the existence of a major antiform in the Pelican ore body, but similar features are not exposed in the Rhinelander area. In Marathon County, a pattern of broad folds with well-developed foliation has been observed by LaBerge and Myers (in prep.). Locally complex deformation in Marathon County is similar to that in areas to the northeast. There is also the possibility of some post-Penokean structures introduced in association with intrusion of the Wolf River batholith (Brown and Greenberg, 1981).

The structural geology of the Penokean volcanic belt can best be summarized as simple folding parallel to the belt axis, followed by faulting and plutonism in a continuous deformation sequence. Increased strain during the Penokean orogeny tightened folds into isoclines and imposed strong axial planar foliations. Local complexity can most often be attributed to stress regimes near plutons and faults in a relationship similar to that observed between plutonism and metamorphism (Fig. 10).

REGIONAL METAMORPHISM WITHIN THE PENOKIAN VOLCANIC BELT

Dutton and Bradley (1970) and Morey (1978) have provided summaries of Precambrian metamorphic patterns in Wisconsin. This work has been updated through recent bedrock mapping (Greenberg and Brown, 1980), which has enabled an improved understanding of the relationship of metamorphism to rock types and structure.

The metamorphism of Penokean rocks can be summarized in terms of contrasting patterns north and south of the Niagara fault. Well-documented metamorphic patterns and mineral assemblages in the northern Penokean terrane are defined by bullseye-type nodes (James, 1955; Cannon, 1973). The nodes have been attributed to basement uplifts and are outlined by annular successions of diagnostic mineral assemblages. The resulting metamorphic zones are well developed in pelitic metasedimentary rocks. However, the relative scarcity of outcrops in general and metapelites in particular south of the Niagara fault have made the definition of metamorphic zones much more difficult in the Penokean volcanic belt. In mafic rocks, such as metabasalts and metaandesites, amphibolite-facies conditions are often

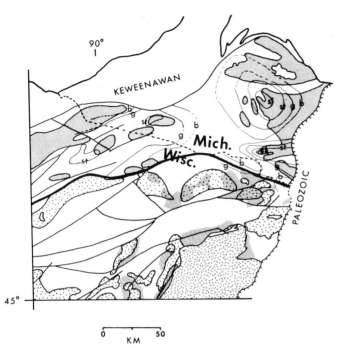

Figure 10. Map of portions of northern Wisconsin and northern Michigan showing nodal metamorphic patterns in the northern Penokean terrane and more localized metamorphism in the Penokean volcanic belt. Major faults are shown in Wisconsin. Archean gneisses and granites are shown by shading. Penokean gneisses and granites are shown by a dotted pattern. The Wolf River batholith granites and syenite are shown by a random dash pattern. Metamorphic isograds are outlined in the northern Penokean terrane. Mineral zones appear as follows: si-sillimanite; st-staurolite; g-garnet; b-biotite. A lined pattern represents areas in the Penokean volcanic belt with amphibolite facies metamorphic assemblages. Modified from James (1955), Dutton and Bradley (1970), Sims (1976), and Greenberg and Brown (1980).

indicated by metamorphic hornblende instead of actinolite, which is stable in most greenschist assemblages. Garnet and biotite were also commonly noted in mafic metatuffs of amphibolite facies. Many metarhyolites exist in areas affected by middle amphibolite-facies metamorphism, but these rocks only show subtle changes, such as recrystallization and chemical-structural transformation of micas, feldspars, and so forth.

It is commonly recognized (Morey, 1978) that the Penokean volcanic belt is predominantly greenschist facies. Rocks of higher metamorphic grade appear to be localized near plutons and in limited areas of intense deformation, such as between plutons or along major structural boundaries (Fig. 10). There is little reason to suspect that this metamorphic intensity correlates with any particular age distinction among Penokean volcanic units. There is also no indication in the Penokean volcanic belt of nodal metamorphic patterns even proximal to Archean gneisses along the southern margin of the volcanic belt (Fig. 1). This contrasts with the nature of Archean basement involvement in the northern Penokean terrane.

Higher grades of metamorphism are developed in deformed granitic rocks of Penokean and Archean age, both within and peripheral to the volcanic belt, respectively. Within the belt, amphibolite-facies metavolcanic and metasedimentary units occur in areas of great structural complexity, such as along the Chippewa River Valley (Myers and others, 1980), in the Mountain area (Lahr, 1972), or in the Rib River area in western Marathon County (LaBerge and Myers, in prep.). Some of these rocks are migmatitic and can be explained by the invasion of magmas or the exposure of deeper levels of erosion, or both.

Within a few kilometres of the northern and eastern periphery of the Wolf River batholith (Fig. 1), staurolite and garnet-bearing assemblages have been attributed to intrusion-related post-Penokean deformation (Brown and Greenberg, 1981; Fig. 10). Some metamorphism related to the intrusion of the Wolf River batholith was characterized by static porphyroblast growth, but in many areas there was also the attendant development of deformational fabrics.

No broad, regional metamorphic patterns are known or would be necessarily expected to accompany the anorogenic 1,760-m.y.-old magmatic event in Wisconsin. There is, however, isotopic and possibly mineralogic evidence for a dynamo-thermal "event" in Wisconsin 1,600 ± m.y. ago (Van Schmus, 1980; Sims and Peterman, 1980; Brandon and others, 1980; Brown and Greenberg, 1981; Geiger and others, 1982). Although direct evidence of 1,600-m.y.-old magmatism is not known in Wisconsin, Brown and Greenberg (1981) have suggested that much localized metamorphism and various complex structural features in central Wisconsin may be related to anorogenic uplift and pluton intrusion at this time. Just how much of the metamorphism and deformation in Wisconsin is actually post-Penokean is difficult to assess at present.

TECTONIC GENESIS OF THE PENOKEAN OROGENY IN WISCONSIN

Because the Penokean volcanic belt of northern Wisconsin is between an Archean craton to the north and some Archean rocks exposed to the south (Fig. 1), different concepts of its evolution must account for the preceding Archean tectonism. These concepts are discussed below and by Larue and Sloss (1980).

The most commonly accepted view of Archean shield development in the southern Lake Superior region is summarized in Sims (1976) and further developed in Sims (1980). In this model, two crustal blocks, a granite-greenstone terrane and an older, northern, gneiss terrane, were welded together in the Archean to form the "basement" for all younger rocks in the region. No other published interpretations have challenged the concept of

Archean tectonics presented in this model. Sims (1976, 1980) also invoked a crustal foundering mechanism to explain the deposition of lower Proterozoic sedimentary and volcanic material in an intracratonal basin upon the Archean crust. The foundering process implies subsidence but not the actual development of rifting. Penokean deformation, metamorphism, and intrusion were attributed to reactivation of the suture between the two Archean terranes.

Although Archean-age rocks are known from only the extreme northwest and southern fringe of the exposed Precambrian in Wisconsin (Fig. 1), it has been suggested that highly deformed and metamorphosed amphibolites within the Penokean volcanic belt are probably Archean (Sims and Peterman, 1980; LaBerge and Myers, in prep.). Sims (1980, p. 115) has also asserted that all gneisses in the region are likely Archean because the Penokean orogeny "was not a gneiss-forming event." If true, this supposition would place specific constraints on tectonic interpretations of the Proterozoic. However, age data (Van Schmus, unpub. data) have shown that at least all the analyzed gneisses from within the Penokean volcanic belt are Penokean, and these obviously indicate "gneiss-forming" conditions after Archean time. As in Wisconsin, investigators once assumed that the lower Proterozoic Flin Flon belt of Manitoba was composed of Archean rocks. In spite of apparent similarities with Archean terranes, dated rock units in the Flin Flon and Penokean volcanic belts are younger.

Another view of the Penokean is based predominantly on radiometric age data and rock distribution (Van Schmus, 1976). In this model, the Penokean is considered as an analogue of Phanerozoic compressive orogenies, with tectonism concentrated along a south-facing margin of the Superior shield. The Archean gneisses south of the Penokean volcanic belt lack the evidence to be considered as inliers of a basement continuous with the Archean craton to the north. More plausible explanations for these exposures are that they may represent rifted continental (cratonic) fragments, or "a microcontinent that was accreted onto the Superior craton during the Penokean Orogeny and formed part of the crust upon which an arc was developed" (Van Schmus and Bickford, 1981, p. 284). A modification of the plate-tectonic idea includes Cambray (1978), which emphasized rifting prior to collisional orogeny along the craton margin. Larue and Sloss (1980) later refined the rifting-sediment interpretations of Cambray (1978). In each case, the concept of collisional orogeny requires a tectonic suture (the Niagara fault) where the Penokean material was welded to the craton. The Niagara fault is also the boundary between the continental margin (northern Penokean terrane) and volcanic arc (Penokean volcanic belt) regimes in the plate-tectonic reconstruction. At present, the fault can be geophysically delineated as the surface separating two crustal blocks of contrasting density (Davis, 1977) and magnetic signature (Zietz and others, 1977).

A most significant contribution to the understanding of the Penokean has been made by Van Schmus and Bickford (1981). They examined these particular rocks along with all other Proterozoic belts in the U.S. midcontinent. The authors discussed several versions of a model to explain the succession of younger magmatic provinces (belts) away from the Archean craton in the north (Fig. 5 in Van Schmus and Bickford, 1981). Their tentative conclusion was that this succession resulted from the progressive accretion and subsequent cratonization of new crust, beginning with the Penokean terranes.

The Penokean material accreted to the Archean craton is similar in many ways to more modern orogenic suites. However, some factors complicate the direct adaptation of a modern plate-tectonic continental margin or island arc model to the Penokean orogeny. As Van Schmus and Bickford (1981) mentioned and in contrast to the interpretation of Upadhyay and Ard (1980), there are no good candidates for ophiolite suites in either Penokean terrane. Known mafic-ultramafic rock complexes in the region are poorly preserved and appear more similar to Archean complexes than modern ophiolites. Penokean metamorphism is also more typical of that in Archean granite-greenstone terranes, where the geometry of metamorphic zones is intimately associated with plutons and local sites of intense deformation (Jolly, 1978). Structural development as a result of the Penokean orogeny is sympathetic with metamorphism, which means that major Penokean structures do not include the horizontal components commonly associated with thrusting and multiphase folding in Phanerozoic orogenies. Vertical tectonic styles are dominant both in Wisconsin (Greenberg and Brown, 1980; Brown and Greenberg, 1981) and in the Upper Peninsula of Michigan (Cannon, 1973; Klasner, 1978). Even along the possible tectonic suture, represented by the Niagara fault (Fig. 1), there is little indication of thrusting (Dutton, 1971: Greenberg and Brown, 1980).

There are problems in directly attributing the Penokean orogeny to simple plate-tectonic processes. However, both Archean and rifting origins for the Penokean volcanic belt are unlikely, and some types of collisional orogeny may be the closest analogue. The structural and metamorphic similarities with tectonism in Archean greenstone belts could indicate that the Penokean event was somewhat transitional in tectonic character between Archean and plate-tectonic orogenies.

Late or post-Penokean age (1,760 m.y. ago; Smith, 1978) volcanism is evident south of the exposed volcanic belt in Wisconsin. The 1,760-m.y.-old volcanic rocks are all rhyolitic and are associated with potassic granites of the same age. The granite-rhyolite suite has been considered as late Penokean, but it is more likely post-Penokean in that the rocks indicate a change to anorogenic tectonism (Anderson and others, 1980). Anorogenic magmatism continued throughout the midcontinent during most of Precambrian time.

SUMMARY AND CONCLUSIONS

Previously uncompiled data from Wisconsin and northern Michigan reveal various aspects of lower Proterozoic volcanism and related tectonism which are considered unique in the midcontinent (Van Schmus and Bickford, 1981). Penokean-age (about 1,900 to 1,850 m.y. ago) volcanic rocks in the Lake Superior region occur in two fundamentally different terranes. Age data suggest that volcanic rocks in the northern Penokean terrane are about 50 m.y. older than those in the Penokean volcanic belt (Van Schmus and Bickford, 1981). The two-terrane tectonic framework is similar to that discussed by Sims (1976; 1980) for the Archean in the same region. The Penokean terranes are separated by the northeast-southwest-trending Niagara fault, which is clearly defined in northeast Wisconsin but can only be traced west by its geophysical signature beneath glacial overburden to the west. The two terranes on either side of the fault represent two distinct tectonic environments whose major contrasts are listed in Table 3.

Penokean sedimentation in the northern terrane indicated increasing tectonic instability and crustal subsidence (Cambray, 1978; Larue and Sloss, 1980). Larue and Sloss (1980) mentioned that the change from platformal clastic and carbonate sediments to turbidites and volcanic materials signaled continental rifting. Prolonged rift development led eventually to a continental margin environment. Volcanic activity in the northern Penokean terrane was dominated by flood basalts and bimodal basalt-rhyolite suites. Chemically, these rocks are equivalent to modern continental-rift suites and some from ocean floor and ridge environments. The calc-alkalic suite from the Emperor Complex (Dann, 1978) may have developed from an Andean-type continental arc on the rifted margin. Descriptions of Penokean-age rocks from Minnesota (Morey, 1972) and the Southern Province of Ontario (Card and others, 1972) make it clear that these areas are continuations of the northern Penokean terrane.

We have shown that volcanic and associated sedimentary rocks in the southern terrane (Penokean volcanic belt) exhibit volcanic arc-derived characteristics. Limited stratigraphic information indicates that depositional sequences in some areas began with basalt-rich lower units, with apparent gradations upward through andesites to more felsic volcanic materials and related clastic sediments. Centers of exhalative activity associated with felsic volcanism are suggested by minor iron formations and massive sulfide mineralization (May, 1977; Cummings, 1978; Bowden, 1978; Schmidt and others, 1978). Sporadic occurrences of

TABLE 3. SPECIFIC CONTRASTS BETWEEN THE TWO PENOKEAN TERRANES

Northern Penokean Terrane	Penokean Volcanic Belt
1) Contains rocks including gneisses dated older than 1860 m.y. (Archean)	1) Contains no rocks dated older than about 1860 m.y.
2) Contains abundant thick sedimentary units and subordinate amounts of tholeiitic volcanic rocks	2) Contains abundant calc-alkalic volcanic rocks and subordinate amounts of diverse sediment types
3) Volcanic rocks dated at 1900 m.y.	3) Volcanic rocks dated at 1850 m.y.
4) Contains major iron formations	4) Contains no major iron formations
5) Contains very few Penokean plutonic rocks	5) Contains large volumes of Penokean plutonic rocks
6) Metamorphism defined by nodal patterns	6) No evidence of nodal metamorphic patterns
7) Contains mantled gneiss domes cored by Archean rocks	7) Contains no mantled gneiss domes

metasedimentary rock types such as graphitic slate, calcsilicate marble, volcanic conglomerate, and slumped turbidites interbedded with flows and tuffs suggest widely variable depositional environments. Observed depositional features such as pillow lavas and sedimentary structures strongly favor subaqueous over subareal environments within the volcanic belt.

The chemistry of volcanic suites south of the Niagara fault are dominantly calc-alkalic. The greater abundance of iron and tholeiitic characteristics noted in Waupee and Quinnesec suites is a reflection of the complex volcanism taking place south of a continental margin. Samples of Waupee volcanic units cannot be classified as alkaline, but they are significantly more alkaline and iron-rich than other suites to the west. Quinnesec Formation samples range into ocean-floor tholeiitic fields on chemical variation diagrams. Pemene Formation samples, also from northeastern Wisconsin, follow an Icelandic trend on the total Fe/MgO diagram (Fig. 6).

Several gabbroic intrusions occurring near the Niagara fault in the northeast Quinnesec area (Prinz, 1959; Dutton, 1971; Greenberg and Brown, 1980) may suggest the possibility of tensional tectonics on a local scale, or they could have intruded along the boundary between Penokean terranes. However, even the more tholeiitic suites south of the fault need not be equated with those in the northern terrane. The large proportion of fragmental volcanic rocks (Garcia, 1978), abundances of andesites (Fig. 4), and common occurrence of chemically intermediate plutons (diorite, tonalite, granodiorite) are orogenic characteristics unique to the southern terrane. These details and the paucity of confirmed Archean basement in the volcanic belt tend to argue against the Penokean basin concept of Sims (1976, 1980).

The Penokean volcanic belt does have a possible analogue in the island arc environments of the modern western Pacific Ocean. Among these arcs, sedimentation is con-

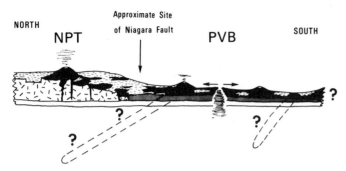

NORTH   Approximate Site
                  of Niagara Fault
   NPT   **PVB**   **SOUTH**

Figure 11. Cross-section of possible plate-tectonic environments for the NPT (northern Penokean terrane) and the PVB (Penokean volcanic belt). Horizontal arrows represent possible interarc spreading. Large dashes and question marks suggest possible subduction. Black areas represent volcanogenic materials of the Andean-type in the NPT and of island arcs in the PVB. Sedimentary materials are represented by a small, dashed pattern. Random dashes represent Archean continental crust. Oceanic crust and the lower crust are represented by a shaded pattern and no pattern, respectively.

trolled by complex, isolated sources including volcanic centers and small interarc basins. Limited interarc spreading and small ocean environments might explain the presence of localized tholeiitic magma generation in the northeastern part of Wisconsin.

It is our conclusion that the best explanation for the origin of Penokean volcanism consists of a rifted continental margin and continental arc that existed north of the proposed suture and a complex island arc situated to the south (Fig. 11). This concept is compatible with previous reconstructions (Van Schmus, 1976; Cambray, 1978; Larue and Sloss, 1980). Existing data limit any more detailed comparisons with modern plate-tectonic environments. We also acknowledge that metamorphic and structural features illustrating the culmination of the Penokean orogeny may not have simple modern analogues. In this one respect the Penokean terranes resemble Archean greenstone belts. However, Penokean volcanic rocks are chemically distinct from Archean magmas and may be products of a primitive style of plate tectonics.

REFERENCES CITED

Anderson, J. L., Cullers, R. L., and Van Schmus, W. R., 1980, Anorogenic metaluminous and peraluminous granite plutonism in the mid-Proterozoic of Wisconsin, U.S.A.: Contributions to Mineralogy and Petrology, v. 74, p. 311–328.

Banks, P. O., and Rebello, D. P., 1969, Zircon ages of Precambrian rhyolite, northeastern Wisconsin: Geological Society of America Bulletin, v. 80, p. 907–910.

Bayley, R. W., and others, 1966, Geology of the Menominee iron-bearing district, Dickinson County, Mighigan, and Florence and Marinette Counties, Wisconsin: U.S. Geological Survey Professional Paper 513, 96 p.

Bougault, H., and Hekinian, R., 1974, Rift valley in the Atlantic Ocean near 36° 50'N: Petrology and geochemistry of basaltic rocks: Earth and Planetary Science Letters, v. 24, p. 249–261.

Bowden, D. R., 1978, Volcanic rocks in the Pelican River massive sulfide deposit, Rhinelander, Wisconsin: A study in wallrock alteration [M.S. thesis]: Houghton, Michigan Technological University, 62 p.

Brandon, C. N., Smith, E. I., and Luther, F. R., 1980, The Precambrian Waterloo quartzite, southeastern Wisconsin: Evolution and significance [abs.]: 26th Institute on Lake Superior Geology, Eau Claire, p. 17.

Brown, B. A. and Greenberg, J. K., 1981, Middle Proterozoic deformation in northern and central Wisconsin [abs.]: 27th Institute on Lake Superior Geology, East Lansing, p. 8.

Burke, K., Dewey, J. F., and Kidd, W.S.F., 1976, Dominance of horizontal movements, arc and microcontinental collisions during the later permobile regime, *in* Windley, B. F., ed., The early history of the Earth: London, Wiley, p. 113–129.

Cain, J. A., 1962, Precambrian granitic complex of northeastern Wisconsin [Ph.D. thesis]: Evanston, Northwestern University, 122 p.

Cambray, F. W., 1978, Plate-tectonics as a model for the environment of sedimentation of the Marquette Supergroup and the subsequent deformation and metamorphism associated with the Penokean orogeny [abs.]: 24th Institute on Lake Superior Geology, Milwaukee, p. 6.

Cannon, W. F., 1973, The Penokean orogeny in northern Michigan, *in* Young, G. M., ed., Huronian stratigraphy and sedimentation: Geological Association of Canada Special Paper 12, p. 251–271.

Card, K. D., and others, 1972, The Southern Province, *in* Price, R. D., ed., Variations in tectonic styles in Canada: Geological Association of Canada Special Paper 11, p. 335–380.

Condie, K. C., 1976a, Trace-element geochemistry of Archean greenstone belts: Earth-Science Reviews, v. 12, p. 393–417.

——1976b, Plate-tectonics and crustal evolution: New York, Pergamon, 288 p.

——1978, Origin and early development of the Earth's crust [abs.], *in* Smith, I.E.M., ed., Proceedings of the 1978 Archean Geochemistry Conference: Toronto, University of Toronto Press, p. 337–338.

Cudzilo, T. F., 1978, Geochemistry of early Proterozoic igneous rocks in northeastern Wisconsin and Upper Michigan [Ph.D. thesis]: Lawrence, University of Kansas, 194 p.

Cummings, M. L., 1978, Metamorphism and mineralization of the Quinnesec Formation, northeastern Wisconsin [Ph.D. thesis]: Madison, University of Wisconsin, 190 p.

Dann, J. C. , 1978, Major-element variation within the Emperor Igneous Complex and the Hemlock and Badwater volcanic Formation [M.S. thesis]: Houghton, Michigan Technological University, 198 p.

Davis, C. B., 1977, Geology of the Quinnesec Formation in southeastern Florence County, Wisconsin [M.S. thesis]: Milwaukee, University of Wisconsin, 133 p.

DeBoer, J., and others, 1980, the Bataan Orogene: Eastward subduction, tectonic rotations, and volcanism in the western Pacific (Philippines): Tectonophysics, v. 67, p. 251–282.

Dutton, C. E., 1971, Geology of the Florence area, Wisconsin and Michigan: U.S. Geological Survey Professional Paper 633, 54 p.

Dutton, C. E., and Bradley, R. E., 1970, Lithologic, geophysical, and mineral commodity maps of Precambrian rocks in Wisconsin: U.S. Geological Survey Miscellaneous Geological Investigation Map I-631.

Edwards, E. C., 1920, The petrography of a portion of Chippewa and Eau Claire Counties, Wisconsin [M.A. thesis]: Madison, University of Wisconsin, 40 p.

Froelich, A. J., 1953, The geology of a part of the Wisconsin granite–Quinnesec greenstone complex, Florence County, Wisconsin [M.S. thesis]: Columbus, Ohio State University, 62 p.

Fulweiler, R. E., 1957, The geology of part of the igneous and metamorphic complex of southeastern Florence County, Wisconsin [M.S. thesis]: Columbus, Ohio State University, 68 p.

Garcia, M. O., 1978, Criteria for the identification of ancient volcanic arcs: Earth-Science Reviews, v. 14, p. 147–165.

Geiger, C., Guidotti, C. V., and Petro, W. L., 1982, Some aspects of the petrologic and tectonic history of the Precambrian rocks at Waterloo, Wisconsin: Geoscience Wisconsin, v. 6, p. 21–40.

Goldich, S. S., and others, 1961, The Precambrian geology and geochronology of Minnesota: Minnesota Geological Survey Bulletin 41, 193 p.

Greathead, C., 1975, The geology and petrochemistry of the greenstone belt, south of Hurley and Upson, Iron County, Wisconsin [M.S. thesis]: Milwaukee, University of Wisconsin, 191 p.

Green, J. C., 1972, Northshore Volcanic Group, *in* Sims, P. K., ed., Geology of Minnesota: A centennial volume: Minnesota Geological Survey, p. 294–331.

Greenberg, J. K., and Brown, B. A., 1980, Preliminary geologic bedrock map of northeastern Wisconsin: Wisconsin Geological and Natural History Survey, scale 1:250,000.

Gunn, B. M., 1976, A comparison of modern and Archean oceanic crust and island-arc petrochemistry, *in* Windley, B. F., ed., The early history of the Earth: London, Wiley, p. 389–403.

Gunn, B. M., Roobol, M. J., and Smith, A. L., 1974, Petrochemistry of the Pelean-type volcanoes of Martinique: Geological Society of America Bulletin, v. 85, p. 1023–1030.

Hall, G. I., 1971, A study of the Precambrian greenstones in northeastern Wisconsin, Marinette County [M.S. thesis]: Milwaukee, University of Wisconsin, 80 p.

Heming, R. F., 1974, Geology and petrology of Rabaul caldera, Papua New Guinea: Geological Society of America Bulletin, v. 85, p. 1253–1264.

Jahn, B. M., and others, 1980, Trace-element geochemistry and petrogenesis of Finnish greenstone belts: Journal of Petrology, v. 21, p. 201–244.

James, H. L., 1955, Zones of regional metamorphism in northern Michigan: Geological Society of America Bulletin, v. 66, p. 1455–1488.

Jenkins, R. A., 1973, The geology of Beecher and Pemene townships, Marinette County, Wisconsin [abs.]: 19th Institute on Lake Superior Geology, p. 15–16.

Jolly, W. T., 1977, Relations between Archean lavas and intrusive bodies of the Abitibi Greenstone Belt, Ontario-Quebec, *in* Baragar, W.R.A., ed., Volcanic regimes in Canada: Geological Association of Canada Special Paper 16, p. 311–330.

——1978, Metamorphic history of the Archean Abitibi Belt, *in* Fraser, J. A., ed., Metamorphism in the Canadian Shield: Geological Survey of Canada Paper 78-10, p. 63–78.

Klasner, J. S., 1978, Penokean deformation and associated metamorphism in the western Marquette Range, northern Michigan: Geological Society of America Bulletin, v. 89, p. 711–722.

Lahr, M. M., 1972, Precambrian geology of a greenstone belt in Oconto County, Wisconsin, and geochemistry of the Waupee metavolcanics [M.S. thesis]: Madison, University of Wisconsin, 62 p.

Larue, D. K., and Sloss, L. L., 1980, Early Proterozoic sedimentary basins of the Lake Superior region: Geological Society of America Bulletin, Part II, v. 91, p. 1836–1874.

Leeman, W. P., and Vitaliano, C. J., 1976, Petrology of McKinney Basalt, Snake River Plain, Idaho: Geological Society of America Bulletin, v. 87, p. 1777–1792.

Maass, R. S., Medaris, L. G., Jr., and Van Schmus, W. R., 1980, Penokean deformation in central Wisconsin: Geological Society of America Special Paper 182, p. 147–157.

Mancuso, J. J., 1960, Stratigraphy and structure of the McCaslin district, Wisconsin [Ph.D. thesis]: East Lansing, Michigan State University, 101 p.

May, E. R., 1977, Flambeau—a Precambrian supergene enriched massive sulfide deposit: Geoscience Wisconsin, v. 1, p. 1–26.

Moore, J. M., 1977, Orogenic volcanism in the Proterozoic of Canada, *in* Baragar, W.R.A., ed., Volcanic regimes in Canada: Geological Association of Canada Special Paper 16, p. 127–148.

Morey, G. B., 1972, General geologic setting of the middle Precambrian, *in* Sims, P. K., ed., Geology of Minnesota: a centennial volume: Minnesota Geological Survey, p. 199–203.

——1978, Metamorphism in the Lake Superior region, U.S.A., and its relation to crustal evolution, *in* Fraser, J. A., ed., Metamorphism in the Canadian Shield: Geological Survey of Canada Paper 78-10, p. 283–314.

Motten, R. H., 1972, The bedrock geology of Thunder Mountain area, Wisconsin [M.S. thesis]: Bowling Green, Bowling Green State University, 59 p.

Muehlburger, W. R., 1980, The shape of North America during the Precambrian, *in* Continental tectonics: Washington, D.C., National Academy of Science, p. 175–183.

Myers, P. E., Cummings, M. L., and Wurdinger, S. R., 1980, Precambrian geology of the Chippewa Valley: 26th Institute on Lake Superior Geology, Guidebook, 123 p.

Prinz, W. C., 1959, Geology of the southern part of the Menominee district, Michigan and Wisconsin: U.S. Geological Survey Open-File Report, 221 p.

Schmidt, P. G., and others, 1978, Geology of Crandon massive sulfide deposit in Wisconsin: Skillings Mining Review, v. 67, no. 18, p. 8–11.

Schriver, G. H., 1973, Petrochemistry of Precambrian greenstones and granodiorites in southeastern Oneida County, Wisconsin [M.S. thesis]: Milwaukee, University of Wisconsin, 83 p.

Schulz, K. J., 1980, The magmatic evolution of the Vermilion greenstone belt, NE Minnesota: Precambrian Research, v. 11, p. 215–245.

Sims, P. K., 1976, Precambrian tectonics and mineral deposits, Lake Superior region: Economic Geology, v. 71, p. 1092–1127.

——1980, Boundary between Archean greenstone and gneiss terranes in northern Wisconsin and Michigan: Geological Society of America Special Paper 182, p. 113–124.

Sims, P. K., and Peterman, Z. E., 1980, Geology and Rb-Sr age of lower Proterozoic granitic rocks, northern Wisconsin: Geological Society of America Special Paper 182, p. 139–146.

Smith, E. I., 1978, Precambrian rhyolites and granites in south-central Wisconsin: Geological Society of America Bulletin, v. 89, p. 875–890.

Stacey, J. S., and others, 1977, Plumbotectonics IIA, Precambrian massive sulfide deposits, *in* Karpenko, S. F., ed., Geochronology and problems of ore formation: U.S. Geological Survey Open-File Report 76-476, p. 93–106.

Stern, R. J., 1979, On the origin of andesite in the northern Mariana island arc: Implications from Agrigan: Contribution to Mineralogy and Petrology, v. 68, p. 207–219.

Thompson, G. L., 1955, A study of part of the Precambrian granite-greenstone complex in southeastern Florence County and northwestern Marinette County, Wisconsin [M.S. thesis]: Columbus, Ohio State University, 86 p.

Turner, F. J., and Verhoogen, J., 1960, Igneous and metamorphic petrology: New York, McGraw-Hill, 694 p.

Upadhyay, H. D., and Ard, K., 1980, Vestiges of Proterozoic back-arc ocean crust in central Wisconsin: Field and petrochemical evidence: Geological Society of America Abstracts with Programs, v. 12, p. 259.

Van Schmus, W. R., 1976, Early and Middle Proterozoic history of the Great Lakes area, North America: Philosophical Transactions of the Royal Society of London, ser. A, v. 280, p. 605–628.

——1980, Chronology of igneous rocks associated with the Penokean orogeny in Wisconsin: Geological Society of America Special Paper 182, p. 159–168.

Van Schmus, W. R., and Bickford, M. E., 1981, Proterozoic chronology and evolution of the mid-continent region; North America, *in*

Kröner, A., ed., Precambrian plate-tectonics: Amsterdam, Elsevier, p. 261–296.

Venditti, A. R., 1973, Petrochemistry of Precambrian rocks in southeastern Oneida County, Wisconsin [M.S. thesis]: Milwaukee, University of Wisconsin, 92 p.

Voight, D. J., 1970, A petrochemical and magnetic study of the volcanic greenstone in northwest Marathon County, Wisconsin [M.S. thesis]: Milwaukee, University of Wisconsin, 94 p.

Weir, K. L., 1967, Geology of the Kelso Junction Quadrangle, Iron County, Michigan: U.S. Geological Survey Bulletin 1226, 47 p.

Whitford, D. J., and Jezek, P. A., 1979, Origin of late-Cenozoic lavas from the Banda arc, Indonesia: Trace-element and Sr isotope evidence: Contributions to Mineralogy and Petrology, v. 68, p. 141–150.

Wisconsin Geological and Natural History Survey, 1981, Bedrock geology of Wisconsin, page-size, scale approximately 1:2,500,000.

Wright, T. L., Grolier, M. J., and Swanson, D. A., 1973, Chemical variation related to the stratigraphy of the Columbia River basalt: Geological Society of America Bulletin, v. 84, p. 371–386.

Zietz, I., Karl, J. H., and Ostrom, M. E., 1977, Aeromagnetic map of Precambrian terrane in Wisconsin: U.S. Geological Survey Miscellaneous Field Studies Map Series, MF-888, scale 1:250,000.

MANUSCRIPT ACCEPTED BY THE SOCIETY MARCH 4, 1983

Geological Society of America
Memoir 160
1983

Early Proterozoic tectonic style in central Wisconsin

R. S. Maass
Department of Geology and Geophysics
University of Wisconsin
Madison, Wisconsin 53706

ABSTRACT

The Penokean orogeny (1,900 to 1,800 m.y. ago) was a major early Proterozoic tectonic event in central Wisconsin, involving deformation, metamorphism, volcanism, and plutonism. Penokean isoclinal F_1 folding affected both Archean and lower Proterozoic rocks, producing a penetrative axial planar foliation that dips steeply throughout the terrane, implying significant horizontal compression during the Penokean orogeny. Foliation has been complexly reoriented by open to tight F_3 folding. Fold axes and penetrative mineral lineation are generally colinear, plunging steeply at all but a few localities. Regional metamorphism accompanying deformation is predominantly lower or middle amphibolite facies.

Penokean tonalite to granite plutons are synkinematic to postkinematic; the earliest plutons intruded during the late stages of F_1 folding. F_1, F_2, and F_3 folding is estimated to have occurred over the interval 1,880 to 1,810 m.y. ago. Although Archean deformation and metamorphism may have affected the terrane, no unequivocal evidence for such events has been recognized. The existence of large-scale Proterozoic shear zones in central Wisconsin is contradicted by evidence which indicates that the proposed zones consist of metavolcanic rocks rather than mylonites.

Spatial separation, differing ages, and contrasting tectonic histories suggest that the Archean gneisses of central Wisconsin do not correlate with the Archean gneisses of Minnesota and northern Michigan.

INTRODUCTION

Although most of the effects of the Penokean orogeny (approximately 1,900 to 1,800 m.y. ago) were recognized a number of years ago in Minnesota (Goldich and others, 1961), northern Michigan (Cannon, 1973), and eastern Ontario (Card and others, 1972), only very recently has the full significance of the Penokean orogeny in Wisconsin become apparent. It is now clear that the Penokean orogeny was a major igneous, metamorphic, and deformational event in central and northern Wisconsin, and it appears that Wisconsin was the site of a greater amount of igneous and deformational activity during the Penokean orogeny than any other area in the Great Lakes region.

The purpose of this paper is to discuss the style, timing, and extent of Penokean deformation in central

Wisconsin. In addition, a metavolcanic interpretation is presented for rocks that previously have been described as mylonites and used as a basis for proposing major Proterozoic shear zones within the state. The approach of the investigation has been to locate and study in detail the best exposures in a terrane that contains extremely limited exposure due to an extensive cover of Paleozoic sedimentary rocks and Pleistocene glacial deposits. The majority of the exposures are located along the banks of the larger streams of the region (Fig. 1). Paucity of outcrop and attendant lack of exposed contact relationships and substantial distances between exposures make it necessary to rely heavily upon radiometric age data to decipher the history of the terrane. Another consequence of sparse ex-

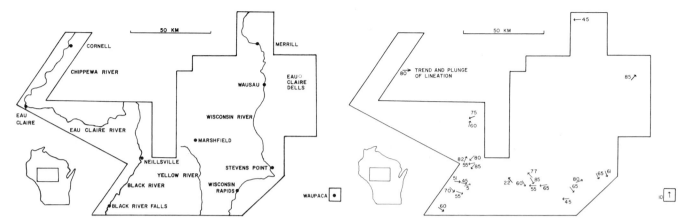

Figure 1. Area of investigation.

Figure 2. Strike and dip of foliation in selected gneisses and foliated plutonic rocks. Plotted strike and dip represents the mean orientation of foliation determined from tens to hundreds of measurements at most localities. This is necessary because the orientation of foliation varies substantially at individual localities due to tight to open F_3 folding of foliation (see Fig. 5). Northern limit of Archean rocks modified from Van Schmus and Bickford (1981).

posure is that macroscopic folds cannot be mapped, and macroscopic fold styles must therefore be inferred from mesoscopic fold styles.

The investigation has focused primarily on the part of the Precambrian terrane south of the dashed line in Figure 2, which consists of an equal abundance of synkinematic to postkinematic Penokean intrusive rocks with Archean gneiss and migmatite. The Archean gneisses are generally finely to thickly banded quartzofeldspathic gneisses containing interlayered amphibolite. Migmatite is relatively rare; only two exposures are known within the terrane. The migmatites consist of banded gneiss intruded by granitic neosomes (Van Schmus and Anderson, 1977).

At many localities the protolith of banded gneiss is not obvious, and it is not clear whether banding represents bedding or is of tectonic-metamorphic origin. Primary sedimentary structures are not present, and composition is commonly tonalitic, allowing tonalite, andesite, or graywacke as possible protoliths. However, along the Black River, the presence of quartzite, banded iron formation, quartz-mica schist, and layers that are interpreted as metatuff and metabasalt indicate sedimentary and volcanic protoliths for some of the gneisses. Many of the gneisses along the Black River are thought to be intermediate and felsic volcanic rocks and associated chert and minor basalt flows.

U-Pb zircon dating of two gneiss and two migmatite localities in the southern portion of the terrane has yielded minimum ages of approximately 2,800 m.y. (Van Schmus and Anderson, 1977; DuBois and Van Schmus, 1978; Van Schmus, 1981, personal commun.). Rb-Sr data from one of the localities led to the suggestion that the protolith might be older than 2,800 m.y. (Van Schmus and Anderson, 1977), but the data can also be explained by Rb loss from a 2,800-m.y.-old protolith during a later metamorphic event (Maass and Van Schmus, 1980). The only Archean plutonic rock recognized is a massive porphyritic syenitic gneiss intruding banded gneiss and yielding a U-Pb zircon age of 2,535 m.y. (Maass and Van Schmus, 1980).

North of the dashed line in Figure 2 the terrane is composed of synkinematic to postkinematic Penokean plutonic rocks, prekinematic or synkinematic Penokean volcanic rocks, and gneisses, which so far have yielded Penokean rather than Archean ages (Van Schmus and Bickford, 1981). The proposed northern limit of Archean rocks in central Wisconsin is based upon limited geochronologic data, and thus the precise location and nature of the Archean–early Proterozoic boundary, if it exists, is unknown. Gneisses on both sides of the proposed boundary are similar enough in terms of lithology, structural style, and metamorphic grade that field criteria for distinguishing between Archean and Penokean gneisses have not been recognized. Penokean plutonic rocks are also similar on both sides of the proposed boundary and approximately equally abundant on both sides.

Penokean plutonic rocks consist of tonalite, granodiorite, granite, and subordinate diorite. They have been intruded by dikes ranging in composition from basalt to rhyolite. Slightly more than half of the plutons and dikes are synkinematic and synmetamorphic; the remainder are postkinematic and postmetamorphic. Units dated by U-Pb zircon methods range in age from 1,875 to 1,820 m.y. (Banks and Cain, 1969; Van Schmus, 1980; Maass and Van Schmus, 1980). Rocks of the same suite to the north of the study area have been dated as 1,885 ± 65 m.y. by Rb-Sr whole-rock techniques (Sims and Peterman, 1980).

Mineral assemblages in mafic dikes, mafic volcanic rocks, and mafic layers and lenses in gneiss provide the major basis for defining the grade of metamorphism in an otherwise quartzofeldspathic terrane almost totally devoid of pelitic, calcareous, and ultramafic rocks. Most of the terrane is lower or middle amphibolite facies as indicated by the assemblage: hornblende-plagioclase (typically An_{25-40}) ± epidote, sphene, biotite, and quartz. Greenschist facies rocks are present in most parts of the terrane but are much less common. Only one upper amphibolite facies occurrence has been identified; it is defined by the assemblage: hornblende-andesine-clinopyroxene-quartz-sphene, in mafic gneiss 30 km north of Eau Claire along the Chippewa River.

STRUCTURAL GEOLOGY

With only a few minor exceptions, the terrane is characterized by steeply dipping banding and foliation (Fig. 2) and steeply plunging fold axes and mineral lineation (Fig. 3). Banded gneisses possess a penetrative foliation accompanied by a penetrative mineral lineation in the plane of foliation. Synkinematic Penokean intrusive rocks possess a penetrative mineral lineation that may or may not be accompanied by a penetrative foliation. Where foliation is present, mineral lineation is in the plane of foliation. Mineral lineation is defined by oriented grains of amphibole and biotite and aggregates of quartz or feldspar. Individual quartz and feldspar grains are generally only slightly elongate parallel to lineation, and many are equant. In rare cases, mostly in the northwestern part of the terrane, mineral lineation is absent from foliated rocks.

Folding in central Wisconsin is separable into two main phases: isoclinal F_1 folding and tight to very open F_3 folding. Tight F_2 folding has been recognized at only a few localities. Since the folds produced during F_2 folding are

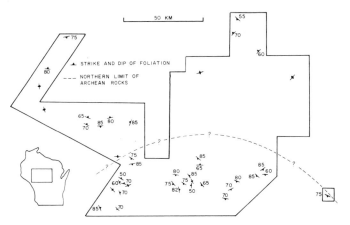

Figure 3. Trend and plunge of lineation at selected localities. Lineation may be mineral lineation, crenulation, or fold axes, or all of the above, since they are generally parallel to each other. Lineation at some localities represents the orientation of F_3 fold axes determined by the stereographic analysis of foliation measurements. The paucity of plotted lineation in the northern part relative to the southern part of the study area is primarily a function of less detailed examination of the northern part of the terrane, but it is also partially due to the fact that mineral lineation is not as well developed in the northern part of the study area.

rare small-scale features, they will not be discussed in detail. For a more complete description of F_2 folding, see Maass and others (1980).

Similar style F_1 isoclinal folding in gneiss produced a penetrative axial planar foliation, S_1, parallel to banding, except in fold hinges where foliation transects banding (Fig. 4). Foliation, where present in Penokean intrusive

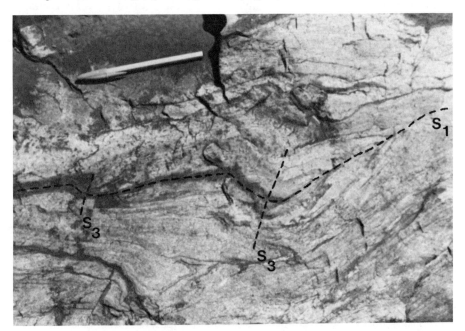

Figure 4. F_1 and F_3 folds in banded gneiss along the Black River. A similar style isoclinal F_1 fold with penetrative axial planar foliation (S_1) has been deformed by open F_3 folds with nonpenetrative axial planes (S_3). Foliation is parallel to banding, except in the hinges of F_1 folds where it transects banding.

rocks, is parallel to foliation in nearby gneiss, and at a locality along the Wisconsin River where the contact between gneiss and a Penokean plutonic rock is exposed, foliation passes through the contact uninterrupted (Anderson, 1972).

No axial planar foliation is associated with F_3 folds (Fig. 4), which are predominantly open in style but locally are tight. Penokean intrusive rocks that were only subjected to F_3 folding possess a penetrative mineral lineation but are unfoliated, with the exception of some dike rocks that possess a penetrative foliation parallel to their margins rather than parallel to regional foliation. Apparently these dikes were mechanically weaker than the surrounding rocks, and thus they preferentially relieved stress by shearing parallel to their margins. Penokean intrusive rocks that were intruded during the final stage of F_3 folding are unfoliated and unlineated but show evidence in thin section of minor granulation and recrystallization.

Along the Wisconsin River, between Stevens Point and Wisconsin Rapids, steeply plunging F_1, F_2, and F_3 fold axes in Archean gneiss are parallel to each other and parallel to penetrative mineral lineation present in Archean gneiss and migmatite and all Penokean intrusive rocks of the area (Maass and others, 1980). These colinear relationships, when taken in conjunction with the parallelism of foliation in Archean gneiss with foliation in Penokean intrusive rocks, indicate that the structures in the rocks of this area were produced during the Penokean orogeny. Detailed study suggests that Penokean intrusive activity began during the waning stages of F_1 folding, and also reveals that the degree of development of foliation can be partially correlated with the age of intrusion (Maass and others, 1980). It is clear that mineral lineation was produced during F_3 folding, since late intrusive rocks that were subjected only to F_3 possess a penetrative mineral lineation, but mineral lineation may also have been forming throughout F_1 and F_2 folding. Since F_1, F_2, and F_3 are coaxial, the question of whether or not mineral lineation was produced during F_1 and F_2 is currently unanswered.

A fold set that does not fit the pattern of Penokean structures along the Wisconsin River is present in nearby Archean migmatite. The axes of isoclinal folds in the migmatite range in plunge from vertical to nearly horizontal. Steeply plunging mineral lineation consistent in orientation with other Penokean linear structures of the area (Maass and others, 1980) transects the hinges of the isoclinal folds, signifying that the folds are older than the mineral lineation. These folds may be the result of inhomogeneous deformation in the migmatite during Penokean F_1 folding, or alternatively, the folds may be Archean.

Along the Black River, it is more difficult to prove that isoclinal F_1 folding occurred during the Penokean orogeny, but structural relationships strongly suggest that this is the case. At Black River Falls, banded gneiss containing F_1 and

F_3 folds and a penetrative mineral lineation is intruded by diabase, diorite, and tonalite, which are similar in character to intrusive rocks of established Penokean age. F_3 fold axes and mineral lineation in the gneiss are parallel to each other and parallel to penetrative mineral lineation in the diabase, diorite, and tonalite. Unfortunately, the orientation of F_1 fold axes cannot be determined at this locality. Ten km to the east, in Algoma-type banded iron formation (Jones, 1978), reconnaissance study indicates that axes of folds with F_1 characteristics are parallel to axes of folds with F_3 characteristics, and both are parallel to penetrative mineral lineation. However, the iron formation is probably of Archean age and, therefore, despite the likelihood that the structures are of Penokean age, they are not conclusively of Penokean age.

Thirty-five km to the northeast of Black River Falls, in the vicinity of Neillsville, F_3 axes and penetrative mineral lineation in Archean gneisses are parallel to each other and to penetrative mineral lineation in nearby Penokean granite and granodiorite. Twenty-five km farther north along the Black River, near Greenwood, mineral lineation in Penokean tonalite is colinear with F_3 axes and mineral lineation in banded gneiss. Foliation in banded gneiss that is axial planar to F_1 isoclinal folds in banded gneiss is parallel to foliation in tonalite, implying that foliation in tonalite is the product of F_1 folding. Tightly folded felsic veins in tonalite have the same orientation of axial planes as isoclinal folds in banded gneiss. Comparison of the two fold styles suggests that tonalite may have been emplaced during the later stages of F_1 folding. As at Black River Falls, the orientation of the axes of F_1 folds could not be determined at either Greenwood or Neillsville.

At three localities along the Black River, axes of F_1 folds are complexly folded, creating diversely oriented interference patterns. The patterns are not consistent with any simple homogeneous folding episode. The explanation for the anomalous patterns becomes apparent at one of the localities where exposure is more extensive than normal. Here it can be seen that the complex folding occurs in the core of a large tight F_3 fold. The complex folding is restricted to the core and is the result of inhomogeneous deformation in the core during F_3 folding.

At Big Falls, 15 km east of Eau Claire along the Eau Claire River, a banded gneiss that formed from a differentiated mafic intrusive (Cummings, 1975) contains isoclinal F_1, tight F_2, and open F_3 folds. A preliminary U-Pb zircon date of the gneiss yielded a Penokean age (Van Schmus, 1979, personal commun.). If this is the age of the protolith, then F_1 through F_3 folding in this part of the terrane is Penokean because a Penokean granite 15 km to the north is undeformed. The other possibility is that the radiometric age of the gneiss is a metamorphic age, in which case folding is still proposed to be Penokean because textural evidence suggests that F_1 axial planar foliation formed during amphibolite facies metamorphism. A nearby, weakly foliated Penokean tonalite also suggests that F_1 is Penokean.

Deformation recorded in rocks along the Yellow and Chippewa rivers is proposed to be of Penokean age, but this remains to be confirmed. Along the Yellow River the style of folding in gneiss is essentially identical to folding along the flanking Wisconsin and Black rivers, but synkinematic Penokean plutonic rocks have not been discovered, and dated gneiss is Archean (Van Schmus and Anderson, 1977). Along the Chippewa River the style of deformation is consistent with that of Penokean age elsewhere, and there are numerous deformed plutonic rocks similar in character to dated synkinematic plutons, but radiometric ages are not yet available.

The strike of foliation is highly variable on a macroscopic scale, as illustrated in Figure 2. The data in Figure 5, collected from one of the largest exposures in the terrane, illustrate the effect of F_3 folding on the orientation of foliation on a mesoscopic scale. Although exposure is too limited to demonstrate it, F_3 folding extrapolated to a macroscopic scale is thought to be the cause of regional

variation. Fold axes, mineral lineation, and crenulation are relatively consistent in orientation on a mesoscopic scale (Fig. 5) but much less so on a macroscopic scale (Fig. 3). Although foliation trends are extremely diverse, it should be noted that east-west to northwest-southeast trends predominate in the southern and western parts of the terrane, in contrast to generally northeast-southwest structural trends reported for northern Wisconsin (Greenberg and Brown, this volume).

The axis of maximum compressive stress during Penokean isoclinal F_1 folding is assumed to be approximately normal to the axial planar foliation produced during this folding event. Since foliation is steep to vertical throughout the terrane, F_1 appears to represent regional horizontal compression, which probably involved substantial horizontal shortening. As discussed earlier, F_3 folding appears to be coaxial with F_1 folding, but the axial plane of F_3 folds is generally at a high angle to the axial plane of F_1 folds. The geometry of F_3 folding appears to require minor to moderate horizontal compression at a high angle to the direction of F_1 compression. Due to the restricted nature of outcrop

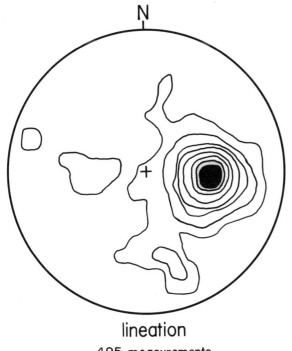

lineation
485 measurements
contours 0.2,1,3,5,10,15,20,25% per 1% area

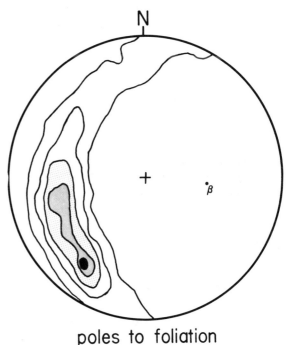

poles to foliation
884 measurements
contours 1,3,5,7,9% per 1% area

Figure 5. Lower hemisphere equal-area projection of structures in the 2,800-m.y.-old Hatfield Gneiss along the Black River (Maass and Van Schmus, 1980). The plot of lineation includes F_1 and F_3 fold axes, mineral lineation, and crenulation, and illustrates the consistent orientation of lineation on a mesoscopic scale. The few anomalous orientations are F_1 fold axes in the inhomogeneously deformed core of a tight F_3 fold (see text). The plot of poles to foliation displays the reorientation of foliation on a mesoscopic scale by tight to open F_3 folding. β defined by the plot of poles to foliation is identical to the mean orientation of lineation. Data collected from a 100×500-m area. Diagrams of this type were used to determine the mean orientation of foliation and lineation at many of the localities plotted in Figures 2 and 3.

Figure 6. Photomicrograph of foliated and lineated Penokean tonalite pluton, showing granoblastic texture typical of deformed and metamorphosed rocks of the terrane. Other rocks of the terrane may exhibit a greater degree of cataclasis, but recrystallization textures resulting from regional amphibolite facies metamorphism are dominant. Major minerals are quartz, andesine, hornblende, and biotite. Foliation trends from upper left to lower right.

and the diverse orientation of structures, the geographic directions of F_1 and F_3 compression are not currently identifiable.

Many of the rocks of the terrane contain cataclastic textures that formed during regional Penokean deformation, but recrystallization dominated, and textures are generally granoblastic (Fig. 6). Textural evidence suggests that cataclasis occurred during amphibolite facies metamorphism, implying a high rate of strain for Penokean deformation. Cataclasis accompanied F_1 and F_3 folding, contributing to the development of both foliation and lineation.

TIMING OF DEFORMATION

The age of Penokean deformation can be roughly estimated because both synkinematic and postkinematic plutons have been radiometrically dated. The age of the onset of deformation is uncertain because the oldest synkinematic plutons were probably emplaced during the late stages of F_1 folding. However, Penokean volcanic rocks appear to be pre-F_1 because their bedding is generally nearly vertical, and these rocks are thought to be no older than an approximately 1,900-m.y.-old rhyolite in the Hemlock Formation of northern Michigan (Van Schmus, 1976). Two volcanic rocks from central Wisconsin and one volcanic rock from northeastern Wisconsin yield ages of 1,860 ± 20 m.y. (Van Schmus, 1980), but these rocks might represent the young end of Penokean volcanic activity. Currently, the age of the volcanic suite in Wisconsin is not well known.

Analysis of two zircon fractions from a synkinematic granite near Neillsville yields a preliminary age of 1,875 ± 25 m.y. (Maass and Van Schmus, 1980). Along the Chippewa, Eau Claire, Black, and Yellow rivers, dated late kinematic to postkinematic plutons indicate that folding ended approximately 1,840 m.y. ago. Along the Wisconsin River, folding commenced before 1,840 m.y. and continued after 1,820 m.y. The present estimate of the age of F_1 through F_3 Penokean folding is 1,880 to 1,810 m.y., but it is open to revision at both ends. No clear spatial-temporal pattern is evident in central Wisconsin, but a larger scale pattern may exist, since Cannon (1973) proposed that Penokean deformation in northern Michigan occurred before 1,900 m.y. ago. Sufficient data are not yet available to test this hypothesis.

EVALUATION OF "SHEAR ZONES" IN WISCONSIN

A number of major "shear zones" have been proposed for central Wisconsin by LaBerge (1972, 1976), LaBerge and Myers (1973, 1977), and Sims and others (1978). The proposed zones have significant tectonic implications, since they are as much as 6 km wide and over 150 km long. In addition, Dutch (1979) has postulated a connection between these zones and the Colorado Lineament described by Warner (1978). Investigation of all of the proposed shear zones in Marathon and Lincoln counties (NE¼ of Fig. 1) has led to the conclusion that the zones consist of metavolcanic rocks rather than mylonites, or in a few cases, synkinematic intrusive rocks flanked by postkinematic intrusive rocks (Maass and Medaris, 1980). Although numerous geophysical lineaments are present in Wisconsin, it is not yet clear what many of these represent, and field evidence fails to support, or in many cases clearly contradicts, a shear zone interpretation of the lineaments. In addition to the conclusion of Maass and Medaris (1980) that the lineaments in Marathon and Lincoln counties are not shear zones, Cummings (1980) stated that field evidence does not support the existence of, or major shearing in, the proposed large-scale Jump River Fault Zone to the northeast of Cornell (Sims and others, 1978).

Detailed examination of the Eau Claire River "shear zone" (LaBerge, 1976) was undertaken at the Eau Claire Dells (Fig. 1) because it is the best exposed and most frequently cited shear zone. The zone is interpreted to consist of a finely to thickly layered sequence of intermediate to felsic tuffs and crystal tuffs, and minor interlayered basalt. The basalt is now amphibolite, and compositionally intermediate units contain abundant garnet. Foliation is parallel to layering, striking N. 35° E. to N. 45° E. and dipping vertically. Steeply plunging mineral lineation is poorly developed in felsic rocks and moderately developed in mafic rocks.

There are many lines of evidence that contradict a mylonite interpretation of the rocks at the Dells. Examination of more than 30 thin sections selected from rocks appearing the most deformed to those appearing the least deformed reveals no significant mortar texture or fluxion structure. The degree of deformation from sample to sample is approximately uniform; there is no progression from undeformed to highly deformed rocks. Megacrysts in the extremely fine-grained groundmass are original quartz and feldspar phenocrysts rather than porphyroclasts, and they are confined to specific layers. Feldspar phenocrysts are generally subhedral and often euhedral (Fig. 7). Quartz phenocrysts are only moderately fractured, and crystal faces and embayments are preserved (Fig. 8). Since quartz is the first mineral to be granulated during cataclasis, this is compelling evidence that the rocks are not mylonites. Foliation wraps around garnet (Fig. 9), indicating that garnet probably grew during the early stage of a compressional metamorphic event. Lack of significant deformation prior to garnet growth is demonstrated by the fact that garnet does not overprint or truncate an early foliation. Lack of shearing during or after garnet growth is suggested by the absence of rotational textures or fracturing of garnet.

Compositional layering on a millimeter scale can be traced across the outcrop with little or no lensing, and the few lensoidal structures that are present do not constitute sufficient evidence for proposing a mylonitic origin of the rocks. As a final argument, the surrounding rocks are not the unsheared equivalents of the rocks at the Dells but are, instead, postkinematic plutons. To the east is the Hogarty Hornblende Granite, a member of the anorogenic 1,485-m.y.-old Wolf River batholith (Van Schmus and others, 1975), and to the west is the 1,825-m.y.-old Kalinke Quartz Monzonite (Van Schmus, 1980). The eastern margin of the Kalinke Quartz Monzonite is sheared (LaBerge and Myers, 1973), but this might represent protoclasis during emplacement.

Figure 7. Photomicrograph of euhedral feldspar phenocryst in felsic crystal tuff from the Eau Claire Dells.

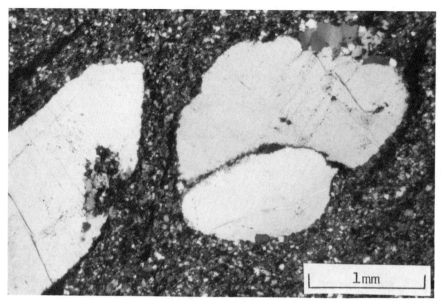

Figure 8. Photomicrograph of quartz phenocrysts in felsic crystal tuff, Eau Claire Dells. Note the preserved crystal faces and embayment texture and the lack of significant fracturing.

The metavolcanic rocks at the Eau Claire Dells share many similarities with early Proterozoic metavolcanic rocks at the Crandon massive sulfide deposit (Schmidt and others, 1978) and the Flambeau massive sulfide deposit (May, 1977). As at the Eau Claire Dells, bedding at the Crandon and Flambeau deposits is nearly vertical. It is thought that the metavolcanic rocks at these three localities formed during the early part of the Penokean orogeny and that the vertical orientation of bedding is due to Penokean F_1 folding.

Throughout the entire area of investigation the degree of deformation is relatively uniform, and there are no recognized zones of unusually intense deformation. As discussed earlier, cataclastic textures are common throughout the terrane, but they are the result of regional Penokean deformation, and their occurrence is not restricted to zones of shearing. At a number of localities, late shearing and faulting has occurred, but it is of relatively minor proportions.

DISCUSSION

In addition to being a major igneous event, the Penokean orogeny was a very significant metamorphic and deformational event in central Wisconsin, affecting both Archean and lower Proterozoic rocks. The parallelism of planar and linear structures in Archean gneiss with planar and linear structures in Penokean rocks strongly suggests that most, if not all, fold structures and fabric in rocks of both ages formed during the Penokean orogeny. The steep to vertical dip of banding and foliation throughout the ter-

rane is interpreted to be the result of substantial horizontal compression during Penokean isoclinal F_1 folding.

Penokean volcanism, occurring primarily in the northern part of the state, but also occurring in central Wisconsin, appears to predate F_1 folding in most cases, but further structural studies are needed to confirm this. Penokean plutonic activity began during the late stages of F_1 folding and continued beyond the termination of F_3 folding. Metamorphism accompanying folding was predominantly lower or middle amphibolite facies; there are local variations to greenschist and upper amphibolite facies.

Although deformation and metamorphism may have occurred during the Archean, no unequivocal structures or textures from an Archean event have been recognized, perhaps due to thorough reworking by Penokean deformation and metamorphism. The available evidence suggests that the Archean was marked primarily by volcanism and sedimentation. Recognized Archean igneous activity consists of the injection of leucosomes into a migmatite in the terrane 2,800 m.y. ago (Van Schmus and Anderson, 1977) and intrusion of a porphyritic syenitic pluton 2,535 m.y. ago (Maass and Van Schmus, 1980).

The tectonic history of Wisconsin contrasts significantly with the tectonic histories of the Minnesota River Valley and northern Michigan. In the Minnesota River Valley, deformation and granulite facies metamorphism occurred during the Archean, producing large, shallow-plunging antiforms and synforms, but the Penokean orogeny was marked only by low-grade metamorphism and minor intrusive activity (Himmelberg and Phinney, 1967; Grant, 1972; Bauer, 1980). During the Penokean orogeny

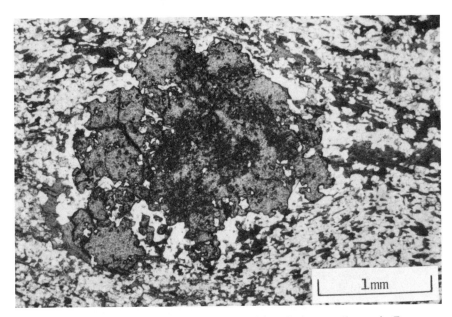

Figure 9. Typical garnet texture in compositionally intermediate unit, Eau Claire Dells. Garnet does not exhibit rotational texture or fracturing. Foliation wrapping around garnet suggests that garnet grew during the early phase of compressional metamorphism.

in northern Michigan, Archean basement was involved in vertical block faulting (Cannon, 1973) and remobilization resulting in gneiss domes (Sims and Peterman, 1976). Metamorphism was low grade, except in nodes reaching sillimanite grade in northern Michigan and extreme northern Wisconsin (James, 1955; Dutton and Bradley, 1970). Very little Penokean plutonic activity has been reported for northern Michigan.

Morey and Sims (1976) and Sims (1976) postulated that the Archean gneiss terranes of Minnesota, northern Michigan, and Wisconsin are part of a single continuous gneiss terrane. However, on the basis of preliminary data, Van Schmus and Bickford (1981) proposed that the gneisses of northern Wisconsin are of early Proterozoic age. According to this model, only the southern margin of the terrane contains gneiss of Archean age (see Fig. 2), leaving a large gap between the Archean gneisses of central Wisconsin and the Archean gneisses of northern Michigan. What lies between the Archean gneisses of central Wisconsin and those of the Minnesota River Valley remains unknown due to Paleozoic cover. Gneisses of the Minnesota River Valley are as old as 3,600 m.y. (Goldich and others, 1980), and some of the gneisses of northern Michigan are at least 3,400 m.y. old, and possibly as old as 3,800 m.y. (Peterman and others, 1980; McCulloch and Wasserburg, 1980). Currently there are no gneisses in central Wisconsin that are known to be older than 2,800 m.y. Because of the differences in age, the dissimilar tectonic histories and styles, and the possibility of a wide intervening early Proterozoic terrane, it appears that the Archean gneisses of

central Wisconsin cannot be correlated with the Archean gneisses of Minnesota and northern Michigan.

Shortly after the Penokean orogeny, Wisconsin once again became the locus of igneous activity; 1,760-m.y.-old granites and rhyolites are present in northern, central, and southern Wisconsin (Van Schmus and others, 1975b; Van Schmus, 1980). Subsequently, the Baraboo and Waterloo quartzites were deposited in southern Wisconsin and then folded along with the rhyolites. Presumably, the granites were also deformed, but little evidence of this exists, perhaps due to a lack of suitable fold markers or the style of deformation. The main criterion for folding in the area is the orientation of bedding in quartzite and rhyolite. In contrast to the style of Penokean deformation, the quartzites, rhyolites, and granites are not penetratively foliated or lineated, although locally the Baraboo quartzite is very weakly foliated, and pelitic layers possess an axial planar cleavage (Riley, 1947). The Baraboo quartzite forms a tight, doubly plunging syncline that has a steeply dipping axial plane striking east-northeast (Dott and Dalziel, 1972). Bedding in rhyolites generally strikes northeasterly and dips steeply (Smith, 1978).

Although recrystallization of the rhyolites has occurred, primary textures are extremely well preserved. Metamorphism of the Baraboo quartzite was very low grade, producing sutured quartz grain boundaries in quartzite and pyrophyllite in pelitic layers (Dott and Dalziel, 1972). To the southeast, in the Waterloo quartzite, pelitic beds reveal that, at least locally, polyphase deformation and metamorphism have occurred, and one of the metamorphic events has reached amphibolite facies (Geiger and others, 1982). The Barron quartzite in northwestern Wisconsin has been correlated with the Baraboo and Waterloo quartzites by Dott and Dalziel (1972). Pelitic layers in the Barron quartzite contain kaolinite and illite, indicating extremely low grade metamorphism.

The event that folded and metamorphosed these rocks is poorly understood, but it is thought to have occurred around 1,630 m.y. ago (Van Schmus and others, 1975b; Smith, 1978). Resetting of Rb-Sr systems in rocks of the lower Great Lakes region took place at this time in what was thought to have been a mild tectono-thermal event (Van Schmus and others, 1975b). Van Schmus and Bickford (1981) have since hypothesized that deformation and metamorphism in Wisconsin at this time is a foreland manifestation of a major event to the south, which has not been recognized due to thick cover of the basement by Paleozoic sediments. The metamorphic effects of this event on central Wisconsin are not yet known, but late epidote, chlorite, and sericite found in many of the rocks of the region may have formed during this event. Evidence of deformation at this time in central and northern Wisconsin has not been recognized, although the Barron quartzite in northern Wisconsin is gently folded (R. H. Dott, Jr., personal commun.). However, the Barron quartzite has not been well studied, and the possibility that it was folded around 1,100 m.y. ago during Keweenawan activity must be considered.

ACKNOWLEDGMENTS

I wish particularly to thank L. G. Medaris, Jr., for suggesting and guiding the investigation. The investigation has benefited immeasurably from the coordinated geochronologic study of the terrane by W. R. Van Schmus. Discussions with W. R. Van Schmus, M. L. Cummings, R. H. Dott, Jr., J. K. Greenberg, B. A. Brown, C. V. Guidotti, W. L. Petro, and J. F. DuBois contributed substantially to my understanding of Wisconsin geology, but I relieve them of all responsibility for the opinions expressed in this report. L. G. Medaris, Jr., J. K. Greenberg, R. L. Bauer, F. W. Cambray, and E. C. Hauser provided constructive criticism of early drafts of the manuscript. Funding was provided by the Wisconsin Alumni Research Foundation (University-Industry Research), the University of Wisconsin Graduate School, the University of Wisconsin Department of Geology and Geophysics, and the Wisconsin Geological and Natural History Survey.

REFERENCES CITED

Anderson, J. L., 1972, Petrologic study of a migmatite-gneiss terrain in central Wisconsin and the effect of biotite-magnetite equilibria on partial melts in the granite system [M.S. thesis]: Madison, University of Wisconsin, 103 p.

Banks, P. O., and Cain, J. A., 1969, Zircon ages of Precambrian granitic rocks, northeastern Wisconsin: Journal of Geology, v. 77, p. 208–222.

Bauer, R. L., 1980, Multiphase deformation in the Granite Falls–Montevideo area, Minnesota River Valley, in Selected studies of Archean gneisses and lower Proterozoic rocks, southern Canadian Shield: Geological Society of America Special Paper 182, p. 1–17.

Cannon, W. F., 1973, The Penokean orogeny in northern Michigan, in Young, G. M., ed., Huronian stratigraphy and sedimentation: Geological Association of Canada Special Paper 12, p. 251–271.

Card, K. D., and others, 1972, The southern province, in Price, R. A., and Douglas, R.J.W., ed., Variations in tectonic styles in Canada: Geological Association of Canada Special Paper 11, p. 381–433.

Cummings, M. L., 1975, Petrology and structure of Precambrian gneisses at Big Falls, Eau Claire County, Wisconsin [M.S. thesis]: Duluth, University of Minnesota.

——— 1980, Preliminary report on the geology of the Jump River Valley: 26th Annual Institute on Lake Superior Geology, Field Trip No. 1 Guidebook, p. 103–111.

Dott, R. H., Jr., and Dalziel, I.W.D., 1972, Age and correlation of the Precambrian Baraboo Quartzite of Wisconsin: Journal of Geology, v. 80, p. 552–568.

DuBois, J. F., and Van Schmus, W. R., 1978, Petrology and geochronology of Archean gneiss in the Lake Arbutus area, west-central Wisconsin [abs.]: 24th Annual Institute on Lake Superior Geology, Milwaukee, Wisconsin, Department of Geology, University of Wisconsin-Milwaukee, p. 11.

Dutch, S. I., 1979, The Colorado Lineament: A middle Precambrian wrench fault system: Discussion: Geological Society of America Bulletin, v. 90, p. 313–314.

Dutton, C. E., and Bradley, R. E., 1970, Lithologic, geophysical, and mineral commodity maps of Precambrian rocks in Wisconsin: U.S. Geological Survey Miscellaneous Geological Investigations Map I-631, scale 1:500,000, with accompanying report, 15 p.

Geiger, C. A., Guidotti, C. V., and Petro, W. L., 1982, Some aspects of the petrologic and tectonic history of the Precambrian rocks of Waterloo, Wisconsin Geoscience Wisconsin, v. 6, p. 21–40.

Goldich, S. S., and others, 1961, The Precambrian geology and geochronology of Minnesota: Minnesota Geological Survey Bulletin 41, 193 p.

——— 1980, Archean rocks of the Granite Falls area, southwestern Minnesota, in Morey, G. B., and others, eds., Selected studies of Archean gneisses and lower Proterozoic rocks, southern Canadian Shield: Geological Society of America Special Paper 182, p. 19–43.

Grant, J. A., 1972, Minnesota River Valley, southwestern Minnesota, in Sims, P. K., and others, eds., Geology of Minnesota—A centennial volume: Minnesota Geological Survey, p. 177–196.

Greenberg, J. K., and Brown, B. A., 1983, Lower Proterozoic volcanic rocks and their setting in the southern Lake Superior district, in Medaris, L. G., Jr., ed., Early Proterozoic geology of the Great Lakes region: Geological Society of America Memoir 160 (this volume).

Himmelberg, G. R., and Phinney, W. C., 1967, Granulite-facies metamorphism, Granite Falls-Montevido area, Minnesota: Journal of Petrology, v. 8, p. 325–348.

James, H. L., 1955, Zones of regional metamorphism in the Precambrian of northern Michigan: Geological Society of America Bulletin, v. 66, p. 1455–1488.

Jones, D. G., 1978, Geology of the iron formation and associated rocks of the Jackson County Iron Mine, Jackson County, Wisconsin [M.S. thesis]: Madison, University of Wisconsin, 117 p.

LaBerge, G. L., 1972, Lineaments and mylonite zones in the Precambrian of northern Wisconsin: Geological Society of America Abstracts with Programs, v. 4, p. 332.

——— 1976, Major structural lineaments in the Precambrian of central Wisconsin, in Hodgson, R. A., and others, eds., Proceedings of the First International Conference on the new basement tectonics: Utah Geological Association Publication No. 5, p. 508–518.

LaBerge, G. L., and Myers, P. E., 1973, Precambrian geology of Marathon County, *in* Guidebook to the Precambrian geology of northeastern and north-central Wisconsin: Wisconsin Geological and Natural History Survey, p. 31–36.
—— 1977, Preliminary geological map of Marathon County, Wisconsin: Wisconsin Geological and Natural History Survey Open-File Report.
Maass, R. S., and Medaris, L. G., Jr., 1980, Metavolcanic rocks at Eau Claire Dells, Marathon County, and an evaluation of the "shear zone" hypothesis in Wisconsin [abs.]: 26th Annual Institute on Lake Superior Geology, Eau Claire, Wisconsin, Department of Geology, University of Wisconsin-Eau Claire.
Maass, R. S., and Van Schmus, W. R., 1980, Precambrian tectonic history of the Black River Valley: 26th Annual Institute on Lake Superior Geology, Field Trip No. 2 Guidebook, 43 p.
Maass, R. S., Medaris, L. G., Jr., and Van Schmus, W. R., 1980, Penokean deformation in central Wisconsin, *in* Morey, G. B., and others, eds., Selected studies of Archean gneisses and lower Proterozoic rocks, southern Canadian Shield: Geological Society of America Special Paper 182, p. 147–157.
May, E. R., 1977, Flambeau—A Precambrian supergene enriched massive sulfide deposit: Geoscience Wisconsin, v. 1, p. 1–26.
McCulloch, M. T., and Wasserburg, G. J., 1980, Early Archean Sm-Nd model ages from a tonalitic gneiss, northern Michigan, *in* Morey, G. B., and others, eds., Selected studies of Archean gneisses and lower Proterozoic rocks, southern Canadian Shield: Geological Society of America Special Paper 182, p. 135–138.
Morey, G. B., and Sims, P. K., 1976, Boundary between two Precambrian W terranes in Minnesota and its geologic significance: Geological Society of America Bulletin, v. 87, p. 141–152.
Peterman, Z. E., Zartman, R. E., Sims, P. K., 1980, Tonalitic gneiss of early Archean age from northern Michigan, *in* Morey, G. B., and others, eds., Selected studies of Archean gneisses and lower Proterozoic rocks, southern Canadian Shield: Geological Society of America Special Paper 182, p. 125–134.
Riley, N. A., 1947, Structural petrology of the Baraboo Quartzite: Journal of Geology, v. 55, p. 453–475.
Schmidt, P. G., and others, 1978, Geologists block out Exxon's big find of Zn-Cu at Crandon: Engineering/Mining Journal, July 1978, p. 61–68.
Sims, P. K., 1976, Precambrian tectonics and mineral deposits, Lake Superior region: Economic Geology, v. 71, p. 1092–1118.
Sims, P. K., and Peterman, Z. E., 1976, Geology and Rb-Sr ages of reactivated Precambrian gneisses and granite in the Marenisco-Watersmeet area, northern Michigan: U.S. Geological Survey Journal of Research, v. 4, p. 405–414.
—— 1980, Geology and Rb-Sr age of lower Proterozoic rocks, northern Wisconsin, *in* Morey, G. B., and others, eds., Selected studies of Archean gneisses and lower Proterozoic rocks, southern Canadian Shield: Geological Society of America Special Paper 182, p. 139–146.
Sims, P. K., Cannon, W. F., and Mudrey, M. G., Jr., 1978, Preliminary geologic map of Precambrian rocks in part of northern Wisconsin: U.S. Geological Survey Open-File Report 78-318.
Smith, E. I., 1978, Precambrian rhyolites and granites in south-central Wisconsin: Field relations and geochemistry: Geological Society of America Bulletin, v. 89, p. 875–890.
Van Schmus, W. R., 1976, Early and middle Proterozoic history of the Great Lakes area, North America: Philosophical Transactions of the Royal Society of London, ser. A, v. 280, p. 605–628.
—— 1980, Chronology of igneous rocks associated with the Penokean orogeny in Wisconsin, *in* Morey, G. B., and others, eds., Selected studies of Archean gneisses and lower Proterozoic rocks, southern Canadian Shield: Geological Society of America Special Paper 182, p. 159–168.
Van Schmus, W. R., and Anderson, J. L., 1977, Gneiss and migmatite of Archean age in the Precambrian basement of central Wisconsin: Geology, v. 5, p. 45–48.
Van Schmus, W. R., and Bickford, M. E., 1981, Proterozoic chronology and evolution of the midcontinent region, North America, *in* Kroner, A., ed., Precambrian plate tectonics: Amsterdam, Elsevier, p. 261–296.
Van Schmus, W. R., Medaris, L. G., Jr., and Banks, P. O., 1975a, Geology and age of the Wolf River batholith, Wisconsin: Geological Society of America Bulletin, v. 86, p. 907–914.
Van Schmus, W. R., Thurman, M. E., and Peterman, Z. E., 1975b, Geology and Rb-Sr chronology of middle Precambrian rocks in eastern and central Wisconsin: Geological Society of America Bulletin, v. 86, p. 1255–1265.
Warner, L. A., 1978, The Colorado Lineament: A middle Precambrian wrench fault system: Geological Society of America Bulletin, v. 89, p. 161–171.

Manuscript Accepted by the Society March 4, 1983

Printed in U.S.A.

Geological Society of America
Memoir 160
1983

Lower Proterozoic stratified rocks and the Penokean orogeny in east-central Minnesota

G. B. Morey
Minnesota Geological Survey
1633 Eustis Street
St. Paul, Minnesota 55108

ABSTRACT

The Animikie basin in Minnesota developed during early Proterozoic time over and approximately parallel to the Great Lakes tectonic zone—an Archean suture between an ancient gneiss terrane to the south and a younger greenstone-granite terrane to the north. The evolution of the basin can be divided into an extensional stage with attendant deposition of stratified rocks, and a subsequent compressional stage assigned to the Penokean orogeny.

The lower Proterozoic stratified rocks in the Animikie basin in Minnesota can be divided into two zones on the basis of pronounced differences in facies and thickness. These are (1) a relatively thin succession of predominantly sedimentary rocks north of the Great Lakes tectonic zone and (2) a much thicker succession of intercalated sedimentary and volcanic rocks to the south. The thicker succession is intensely deformed, metamorphosed, and intruded locally by igneous rocks. The close correspondence between the sedimentological and tectonic patterns implies that sedimentation and tectonism were both part of a tectonic continuum that began with the development of the depositional basin and culminated with the major tectono-thermal pulse of the Penokean orogeny. If a tectonic continuum is assumed, then any hypothesis to account for extension in the basin must also account for the compression.

A close correspondence between sedimentological and tectonic patterns exists between the lower Proterozoic strata in Minnesota and those of the southeastern segment of the Animikie basin in Wisconsin and northern Michigan. Therefore, it seems likely that sedimentation in both segments occurred within similar tectonic regimes and that any hypothesis to account for the evolution of either segment must account for the evolution of both.

The similarity of sedimentological attributes of lower Proterozoic strata in the Animikie basin and Phanerozoic strata deposited in geosynclines has prompted various plate-tectonic models to explain the formation of the basin. The structural features formed in the basin during the compressional stage, however, lack attributes typical of a consuming continental margin or colliding continental plates. The lack of these attributes does not negate the usefulness of plate-tectonic processes in interpreting the basin's history, nor does it prove that such processes were not operative. We still do not know if the evolution of the Animikie basin followed the same "rules" of extension and compression that govern the evolution of a "typical" Phanerozoic geosyncline, or if the basin formed by a sequence of processes unique to its time and place.

INTRODUCTION

The rocks of the Animikie basin have been extensively studied by several generations of geologists, mainly because they contain the vast iron-formations for which the Lake Superior region is famed. Because of their economic importance, most of the iron-bearing districts have been mapped in detail.

During the past 20 years, significant progress has been made in understanding the Proterozoic geology of the Lake Superior region. This is particularly true of the lower Proterozoic rocks (Proterozoic X of U.S. Geological Survey usage) that occur in a broad intracontinental basin, which underlies much of east-central and northern Minnesota, adjacent parts of Ontario, northern Wisconsin, and the northern peninsula of Michigan. This basin has been informally termed the Animikie basin (e.g., Trendall, 1968; Sims, 1976) because most strata in it were once assigned to the so-called Animikie Series of James (1958). The term "Animikie Series" has now been abandoned as a lithostratigraphic term, but the term "Animikie basin" has been retained for convenience.

The radiometric studies of Goldich and others (1961), Aldrich and others (1965), Goldich (1968), Van Schmus (1976, 1980), and numerous others provide a chronometric framework within which both intradistrict and interdistrict correlations can be made. Accordingly, there has been a gradual shift from descriptions of individual mining districts to syntheses of broader geologic environments, such as those by Morey (1973, 1979), Bayley and James (1973), Sims (1976), LaRue and Sloss (1980), Sims and others (1981), and Sims and Peterman (1983). The various published syntheses have led to a much better understanding of the regional attributes of the Animikie basin, but there is still some controversy concerning the tectonic processes responsible for the deposition and subsequent deformation of the lower Proterozoic rocks within the region. In essence, the controversy revolves around the question of whether the rocks are the product of Phanerozoic-like plate-tectonic processes (Van Schmus, 1976; Cambray, 1977, 1978a, 1978b; LaRue and Sloss, 1980) or of sedimentation and deformation in an intracontinental setting unique to early Proterozoic time in the Lake Superior region (Sims, 1976; Morey, 1979; Sims and others, 1980). Both points of view have been based primarily on geologic observations in Upper Michigan and adjoining parts of northern Wisconsin because until recently the stratigraphic history of correlative rocks in Minnesota was poorly understood. However, with the recent completion of a detailed geologic map (Morey and others, 1981), the basic geologic features of the Animikie basin in Minnesota are now established. More sophisticated research on the sedimentary and metamorphic petrology and the structure of the rocks in the basin needs to be done, but there now are sufficient data to compare the evolutionary history of the Animikie basin in Minnesota with that of the basin in northern Wisconsin and Upper Michigan.

REGIONAL GEOLOGIC SETTING

In terms of surface exposures the Animikie basin occurs at the southern extremity of the Canadian Shield where it forms a major part of the so-called Southern province (Stockwell and others, 1970) or the Hudsonian foldbelt on the *Tectonic Map of North America* (King, 1969). However, if the subsurface geology is considered, the basin occurs near the center of the known Precambrian basement of the North American craton (Fig. 1).

Archean rocks of two contrasting terranes, which differ in age, rock assemblages, metamorphic grade, and structural style (Morey and Sims, 1976), form the basement for the supracrustal rocks of the Animikie basin (Fig. 2). Greenstone-granite complexes of late Archean (2,750 to 2,600 m.y.) age, which are typical of most of the southern part of the Superior province, underlie the northern part of the basin. In contrast, migmatitic gneiss and amphibolite, in part about 3,600 m.y. old, underlie the southern part of the basin (Peterman, 1979). The type area for the gneiss

Figure 1. Generalized map of the North American continent showing the location of the Lake Superior region relative to the Canadian Shield and to known or inferred Precambrian basement rocks of the North American craton (modified from Shoemaker, 1975).

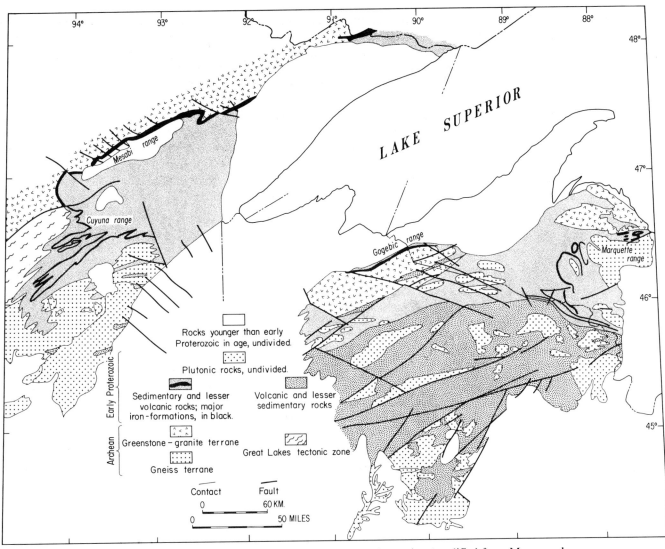

Figure 2. Generalized geologic map of the Lake Superior region (modified from Morey and others, 1982) showing the relationship of the Animikie basin to Archean rocks of the gneiss terrane, the greenstone-granite terrane, and the Great Lakes tectonic zone. Not shown are the middle Proterozoic Keweenawan rocks of the Midcontinent Rift System that divide the Animikie basin into two discrete segments.

terrane is the Minnesota River Valley in southwestern Minnesota; the type area for the greenstone-granite terrane is the Vermilion district of northeastern Minnesota (Sims, 1980). The two Archean terranes are juxtaposed along a major crustal feature—the Great Lakes tectonic zone of Sims and others (1980)—which extends more than 1,200 km between central South Dakota and the Grenville front in eastern Ontario. The zone formed in late Archean time when the two crustal segments were joined into a single, large continental block (Sims and Morey, 1973). Although the two segments were juxtaposed thereafter, the boundary zone was the focus of major crustal movements during much of early Proterozoic time.

As judged from chronometric evidence, the Animikie basin developed over and along both sides of the Great

Lakes tectonic zone during the period 2,100 to 1,850 m.y. ago (Beck and Murthy, 1982; Van Schmus and Bickford, 1981). Most of the rocks within the Animikie basin are of clastic origin, but the basin also contains nearly all of the commercially exploited iron-formations of the Lake Superior region as well as appreciable volumes of volcanic rocks, particularly in Upper Michigan.

An additional, possibly coeval sequence (Van Schmus, 1976, 1980) of dominantly mafic volcanic rocks occurs in a number of possibly fault-bounded basins in northeastern, north-central, and south-central Wisconsin (Sims and Peterman, 1983) and as small isolated roof pendants and septa within and between granitic plutons in east-central Minnesota (Morey and others, 1982). Rocks of this sequence are poorly exposed, and they are nowhere directly associated with strata of the Animikie basin. Consequently, there is no definitive geologic evidence as to the stratigraphic position of this volcanic sequence relative to the better known succession in the basin proper (Sims and Peterman, 1983).

Sedimentation and volcanism within and to the south of the Animikie basin were terminated or closely followed by a tectono-thermal event, the Penokean orogeny of Goldich and others (1961), which occurred mainly about 1,900 to 1,800 m.y. ago (Van Schmus, 1980). The resulting deformation and metamorphism of the supracrustal rocks were most intense over the gneiss terrane where the gneisses make up the cores of domes or fault-bounded antiformal blocks surrounded by superjacent bedded rocks of early Proterozoic age (Cannon, 1973; Sims, 1976, 1980; Morey and Sims, 1976; Sims and others, 1980). A second distinctive feature of the Penokean orogeny was the formation of concentric, nodal patterns of metamorphism that reached the staurolite and locally the sillimanite grade (James, 1955) around the flanks of the up-faulted blocks or domes of gneissic basement rock (Sims, 1976).

Toward the end of the Penokean orogeny and mainly after its termination, plutons of generally calc-alkalic to alkalic affinity, ranging in composition from tonalite or diorite to granite, intruded the stratified rocks, particularly in east-central Minnesota (Morey, 1978) and northeastern Wisconsin (Bayley and others, 1966). In general, the older and more mafic plutons are somewhat deformed or sheared, whereas the younger granitic plutons are generally undeformed (Sims and Peterman, 1980; Van Schmus, 1980).

The lower Proterozoic rocks are overlain and intruded by a variety of rocks that range in age from approximately 1,760 m.y. (Dott and Dalziel, 1972; Van Schmus and others, 1975; Van Schmus, 1978, 1980; Smith, 1978) to the middle part of Late Cretaceous time.

The Midcontinent Rift System, and its associated Keweenawan mafic igneous and sedimentary rocks deposited about 1,200 to 1,100 m.y. ago, divides the Animikie basin into two isolated segments. Consequently, the strata in the northwestern segment are assigned to the Animikie and Mille Lacs Groups (Marsden, 1972; Morey, 1973, 1978), and those in the southeastern segment are assigned to the Marquette Range Supergroup (Cannon and Gair, 1970). The Marquette Range Supergroup is much thicker and more diverse than the sequence in Minnesota (Fig. 3) and is interrupted by unconformities, which serve to divide it into the Chocolay, Menominee, Baraga, and Paint River Groups (James, 1958, p. 30). Nonetheless, the Animikie and Menominee Groups and the Mille Lacs and Chocolay Groups are lithically quite similar, and this similarity provides much of the evidence for suggested correlations of strata between the northwestern and southeastern segments (James, 1958; Bayley and James, 1973).

GEOLOGY OF THE NORTHWESTERN SEGMENT OF THE ANIMIKIE BASIN

The lower Proterozoic stratified rocks in the northwestern segment of the Animikie basin form a broad, east-northeast-plunging synclinorium bounded on the

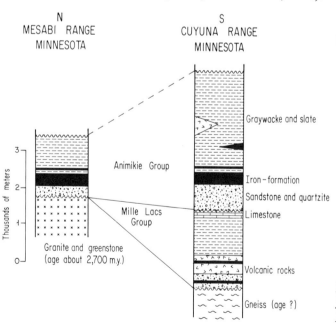

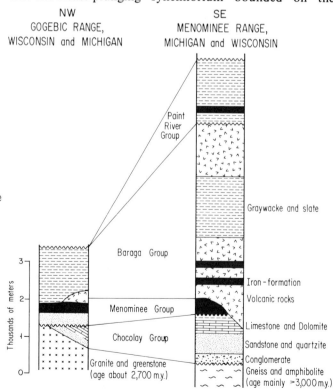

Figure 3. Generalized correlations of lower Proterozoic strata in the northwestern segment (Morey, 1978) and in the Marquette Range Supergroup in the southeastern segment (Cannon and Gair, 1970) of the Animikie basin.

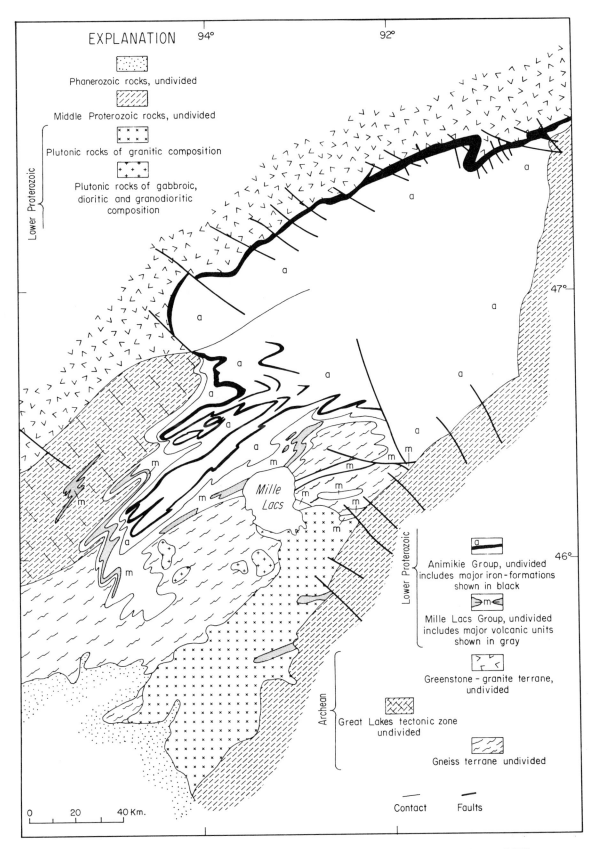

Figure 4. Generalized geologic map of east-central Minnesota modified from Morey (1979) and Morey and others (1981). See text for discussion of units.

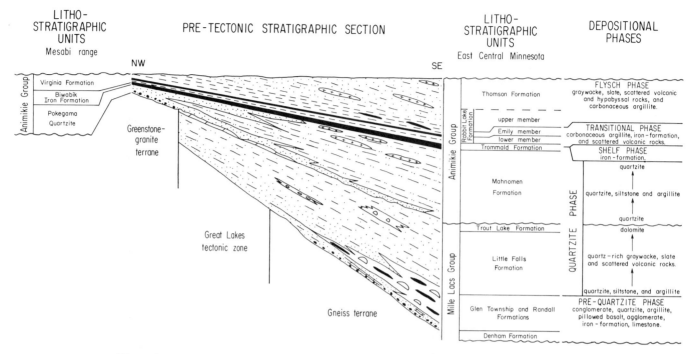

Figure 5. Pretectonic cross section showing the relationship between lithostratigraphic units and depositional phases of the lower Proterozoic stratified rocks in east-central Minnesota (modified from Morey, 1979).

north, west, and southeast by Archean rocks (Fig. 4). Where rocks of the Animikie Group unconformably overlie Archean greenstone and granite on the northern edge of the basin, they are at least 2 km thick (Chandler and others, 1982). However, to the south, and particularly over the gneiss terrane, the Animikie Group is 2 to 3 km thick (Chandler and others, 1982) and is underlain by the Mille Lacs Group, a lower Proterozoic sequence, which is at least several kilometers thick (Morey, 1978).

Stratigraphy

As defined by Morey (1978), the Denham Formation at the base of the Mille Lacs Group (Fig. 5) consists dominantly of quartz-rich sandstone of arenitic affinity, and lesser amounts of conglomerate, dolomite, oxide-facies iron-formation, and subaqueous volcanogenic rocks of mafic to intermediate composition. In several places the Denham Formation passes laterally into areally extensive, thick units consisting dominantly of mafic to intermediate subaqueous volcanic rocks intercalated with lesser amounts of carbonate-facies iron-formation (Han, 1968), pyrite-rich carbonaceous argillite, and quartz-rich arenites; these dominantly volcanic sequences are assigned to the Glen Township and Randall Formations (Fig. 5). All of the basal units pass gradationally upward into the Little Falls Formation, a thick sequence of interbedded quartz-rich wacke, siltstone, shale, and only minor amounts of volcanogenic

rocks. Lenses and beds of impure dolomite or limestone also are present throughout the Little Falls Formation and are particularly abundant in the upper part of the Mille Lacs Group where they compose a mappable unit named the Trout Lake Formation (Marsden, 1972).

An unconformity may separate rocks of the Mille Lacs Group from those of the overlying Animikie Group on the Cuyuna range (Marsden, 1972). The Animikie Group there and on the Mesabi range to the north consists of a basal, well-sorted, epiclastic sequence assigned to the Mahnomen Formation or the Pokegama Quartzite. The clastic rocks are for the most part abruptly overlain by iron-rich strata of the Trommald or Biwabik Formations, which in turn are gradationally overlain by thick turbidite sequences assigned to the Rabbit Lake, Thomson, or Virginia Formations (Morey, 1973).

The well-sorted clastic detritus of the Pokegama Quartzite ranges in thickness from no more than a few meters at the eastern end of the Mesabi range to as much as 100 m at the westernmost end of the Mesabi range. Correlative strata of the Mahnomen Formation on the Cuyuna range to the south are at least 600 m thick (Schmidt, 1963); therefore, subsidence during Pokegama-Mahnomen time appears to have been greater in the southern part of the basin.

The iron-rich strata of the overlying Biwabik or Trommald Formations generally are 100 to 200 m thick and are characterized by intercalated facies indicative of

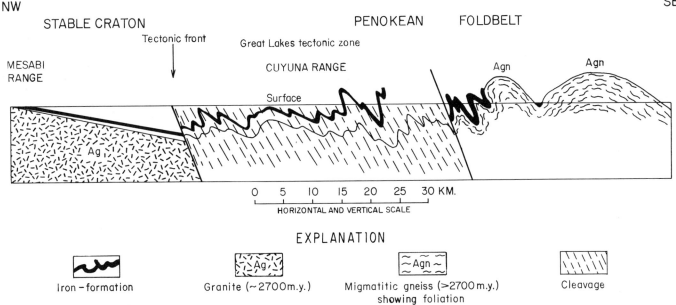

Figure 6. Geologic section across the Animikie basin in Minnesota. Modified from Morey (1978) and Sims and others (1980).

both shallow and "deeper water deposition" (Morey, 1973). Volcanic rocks, mainly of pyroclastic origin, are sparingly present on the Mesabi range (French, 1968), but pyroclastic and extrusive volcanic rocks occur commonly in the Trommald Formation on the Cuyuna range (Schmidt, 1958).

Overlying the iron-formations is more than 900 m of intercalated carbonaceous mudstone and siltstone assigned to the Virginia Formation on the Mesabi range and to the Rabbit Lake and Thomson Formations in east-central Minnesota. The lower 60 m or so of this depositional phase is characterized by black, carbonaceous shale and siltstone, iron-formation (e.g., the Emily Iron-formation Member of the Rabbit Lake Formation on the Cuyuna range), and igneous material including pyroclastic material, flows, and thin hypabyssal dikes or sills. Much of this sequence, however, contains appreciable quantities of graywacke (Morey, 1969; Morey and Ojakangas, 1970; Lucente and Morey, 1983), as well as beds of carbonaceous sulfide-facies iron-formation, apatite, chert, felsic and mafic tuff, mafic and andesitic lava flows, and coeval diabasic intrusions (Morey, 1978; McSwiggen, 1982; Connolly, 1981).

Deformation and Metamorphism Associated with the Penokean Orogeny

The stratified rocks of the Mille Lacs and Animikie Groups were extensively deformed and metamorphosed during the Penokean orogeny at about 1,870 m.y. (Peterman, 1966; Keighin and others, 1972). The degree of de-

formation, which varies considerably from north to south, appears to be related to the contrasting kinds of underlying Archean rocks. The stratified rocks can be divided into two broad longitudinal zones (Fig. 6) on the basis of contrasting styles of deformation and grades of metamorphism (Sims, 1976)—a northern stable cratonic zone and a southern deformed zone termed the Penokean foldbelt (Sims and others, 1980). The tectonic front separating the two zones coincides with the inferred northern edge of the Great Lakes tectonic zone (Sims and others, 1980) and is marked in east-central Minnesota by the northern limit of a penetrative cleavage (Marsden, 1972; Morey, 1978).

North of the tectonic front, the supracrustal rocks, which unconformably overlie Archean greenstone-granite complexes, were virtually undeformed and unmetamorphosed during the Penokean orogeny. Strata on the Mesabi range dip gently southward, and the surface of the basement rocks appears to be relatively undisturbed except for block faulting. In contrast, rocks of the Penokean foldbelt over the Great Lakes tectonic zone are significantly metamorphosed, and both the supracrustal rocks and the underlying basement rocks appear to be complexly infolded. This part of the Penokean foldbelt is characterized by several contrasting tectonic styles, which differ from north to south, partly because different structural levels are exposed (Fig. 6). The northernmost subzone, immediately south of the tectonic front, is characterized by several large anticlines and synclines with many coaxial second- and third-order folds on their limbs (Marsden, 1972). The folds

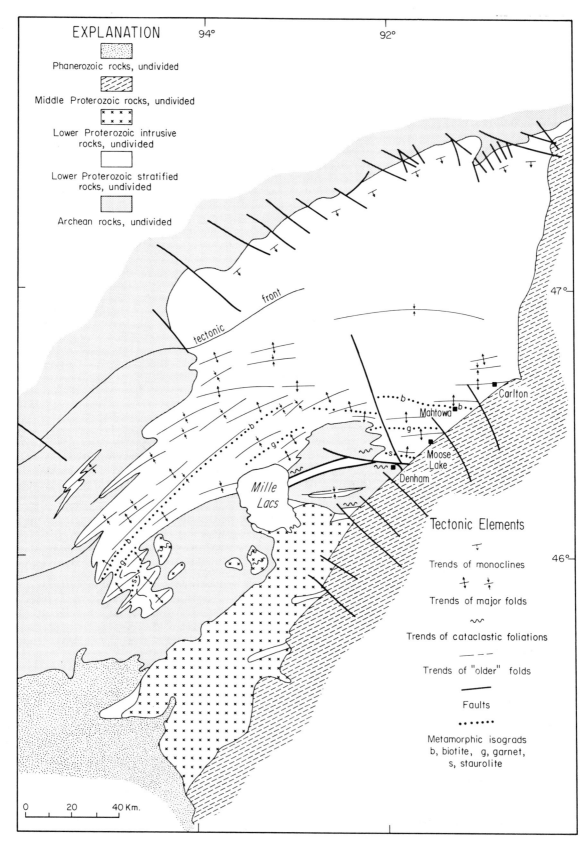

Figure 7. Tectonic map showing major fold axes and metamorphic mineral isograds related to the Penokean orogeny in east-central Minnesota. Metamorphic data modified from Morey (1979).

have nearly vertical, straight to broadly curvilinear axial planes that trend east-northeast (Fig. 7). The southern subzone is characterized by folds that are steeply overturned to the northwest and have a steep southeast-dipping penetrative cleavage (Schmidt, 1963) or a discontinuous crenulation cleavage, which developed as an axial plane foliation in rocks of higher metamorphic grade (Connolly, 1981).

Deformation in the Penokean foldbelt just to the south of the southern edge of the Great Lakes tectonic zone is manifested principally by several domes, which have cores of gneiss and are locally overlain by steeply dipping stratified rocks. The cores and particularly their margins are indented in places by infolds of stratified rocks that are oriented in a generally east-northeast direction. The gneissic rocks near core-mantle boundaries are sheared. The cataclastic zones are marked by thin plates of oriented biotite and locally by hornblende aggregates and boudins that are subconformable to the bedding and other structural elements in the stratified rocks that overlie and surround the domes (Keighin and others, 1972).

Although generally east-trending folds dominate map patterns in the Penokean foldbelt (Fig. 7), there is considerable evidence on an outcrop scale for an earlier period of folding in the Thomson Formation (Connolly, 1981) where overturned or recumbent isoclinal folds have axes that plunge at low angles to the east and west and axial planes that are horizontal or dip at low angles to the south. A metamorphic schistosity, which parallels bedding in most outcrops, developed during this deformation as an axial-plane foliation to these isoclinal folds.

Structural elements mapped at the southwestern end of the basin include an apparent earlier generation of folds refolded about northeast-trending axes (Fig. 7). Although these early structures represent a period of folding prior to the main phase of the Penokean orogeny, their original geometry and temporal relationship to the early folds in the Thomson Formation are unknown.

The east-trending folds that dominate the Penokean foldbelt have also been refolded about north-northwest-trending axes in parts of the Cuyuna range. Although this younger fold system has been worked out in only one or two small open-pit mines, it may extend over much of the Cuyuna range (Schmidt, 1959). Regardless, the lower Proterozoic stratified rocks have been affected by three and possibly four periods of folding.

Metamorphic patterns in the lower Proterozoic strata mimic the major structural pattern, that is, the metamorphic grade increases from north to south (Fig. 7). North of the tectonic front, the stratified rocks contain mineral assemblages indicative of metamorphic conditions scarcely above diagenesis (Morey, 1973; Perry and others, 1973; Floran and Papike, 1975; Lucente and Morey, 1983). Argillaceous rocks, which occur from immediately south of the tectonic front to north of Mahtowa, contain mineral as-

semblages indicative of the lower greenschist facies (Marsden, 1972). Near Mahtowa the argillaceous beds are phyllitic, and intercalated metagraywacke beds are recrystallized sufficiently to obliterate most primary textures. Biotite first appears in argillaceous rocks just south of Mahtowa where they and the interbedded metagraywacke units have a pronounced schistosity. Garnet appears as a prograde mineral phase approximately 14 km south-southwest of Mahtowa, where carbonate concretions contain scattered grains of hornblende (Weiblen, 1964). The supracrustal rocks near Denham have been metamorphosed to the lower amphibolite facies; garnet occurs in more quartzose beds, hornblende in calcareous and volcanic units, and staurolite in argillaceous units. Adjacent gneissic basement rocks also have been metamorphosed to the lower amphibolite facies as evidenced by secondary hornblende and K-feldspar, particularly within cataclastic zones (Keighin and others, 1972; Morey, 1978).

Closely spaced biotite, garnet, and staurolite isograds have also been mapped in the lower Proterozoic stratified rocks at the southwest end of the basin (Fig. 7). Between this area and Denham, however, the isograds shown on Figure 7 are merely inferred from scattered outcrops and drill holes.

Penokean and Post-Penokean Igneous Rocks

Lower Proterozoic plutonic rocks are confined to that part of east-central Minnesota presumably underlain by Archean gneisses. The wide variety of inclusions of metavolcanic and metasedimentary origin in the plutonic rocks, as well as the presence of metavolcanic rocks as septa between several plutons and between plutons and basement rocks, indicates that igneous activity occurred in an area once overlain by lower Proterozoic stratified rocks.

Igneous activity was characterized by the emplacement of several discrete intrusive bodies of variable size (Fig. 4) and of calc-alkalic to alkalic affinity (Morey, 1978). Because these intrusions are in sharp contact with one another and contain inclusions of other rock types, it is possible to define a sequence of intrusive events. Dikes, sills, and stocks of gabbroic to dioritic composition, too small to be shown on Figure 4, are inferred to be the oldest lower Proterozoic plutonic rocks in east-central Minnesota. Most of these intrusions occur along cleavage planes or at intersections of joints or other fractures in the stratified rocks. They are especially common at the southwestern end of the Penokean foldbelt. Although crosscutting relationships have not been seen, it is inferred that this period of igneous activity was followed by the emplacement of several small stocks of granodioritic composition and later by the emplacement of several large granitic plutons of generally sodic composition. Plutonic igneous activity culminated with the emplacement of a composite batholithic unit consisting of several kinds of potassic granitic rocks as well as a discrete border facies with rapakivi-like textures.

DISCUSSION

The lower Proterozoic stratified rocks in the north-western segment of the Animikie basin can be divided into two zones on the basis of pronounced differences in facies and thickness. These are (1) a relatively thin succession of predominantly sedimentary rocks in the north and (2) a much thicker succession of intercalated sedimentary and volcanic rocks in the south. The thicker succession is intensely deformed, metamorphosed, and intruded locally by igneous rocks. The close correspondence between the sedimentological and tectonic patterns implies that sedimentation and tectonism were both part of a tectonic continuum that began with the development of the depositional basin and culminated with the major tectono-thermal pulse of the Penokean orogeny. Thus the history of the Animikie basin can be divided into an early extensional stage with attendant deposition and a subsequent compressional stage assigned to the Penokean orogeny.

Extensional Stage

Most sedimentological models for the Animikie basin rely on the work of James (1954) and Bayley and James (1973) who concluded that the Marquette Range Supergroup in Michigan records a complete transition from "stable craton" to a eugeosynclinal environment during early Proterozoic time. Evidence for such a transition is especially clear in Minnesota where the sedimentary record can be divided into five depositional phases (Fig. 5; Morey, 1979). The pre-quartzite and quartzite phases constitute a miogeosynclinal sequence during which sediments were derived from both north and south of the basin. The third phase represents a shelf upon which iron-rich strata were precipitated, whereas the fourth phase forms a transitional sequence marked by rapid subsidence of the shelf and concurrent deposition of black, carbonaceous muds. In the flysch phase a southward-thickening, clastic wedge was deposited by southward-flowing turbidity currents associated with a number of submarine fan complexes (Lucente and Morey, 1983).

Rocks of the earliest or pre-quartzite phase include a discontinuous veneer of coarse- to fine-grained, generally well sorted epiclastic rocks and oxide-facies iron-formation that passes laterally into a coeval complex of pillowed basalt, agglomerate, carbonate-facies iron-formation, and black carbonaceous slate. The epiclastic sequence consisted originally of quartz-rich arenite and mudstone and smaller amounts of conglomerate and limestone or dolomite. The conglomerate occurs near the base of the sequence as lens-like sedimentation units that are massive to very poorly bedded. They contain pebble-size and larger clasts that appear to have been locally derived from subjacent gneisses. The limestone and dolomite occur as beds in the argillaceous parts of the sequence; they consist generally of hypidiomorphic grains of calcite, some of which enclose scattered grains of detrital quartz and feldspar.

Mafic volcanic rocks of the pre-quartzite phase are characterized by pillow structures indicative of subaqueous deposition. In places they are interlayered with beds of agglomerate, chert, iron-formation, and sulfide-bearing carbonaceous slate. In places the sequence is intruded by dikes or sills of diabasic gabbro that likely are the hypabyssal equivalents of the basalts.

The epiclastic rocks of the pre-quartzite phase (Fig. 8A) apparently were deposited in a shallow-water, fairly high energy depositional regime under fluvial to possibly marginal marine conditions. In contrast, the volcanic complex and its associated sedimentary rocks were deposited under anoxygenic conditions and therefore probably in deeper water. The transition from the thin, dominantly epiclastic sequence to the thick, dominantly volcanic sequence is apparently abrupt and occurs approximately over the inferred southern edge of the Great Lakes tectonic zone. This relationship implies that the volcanic and associated deeper rocks were deposited in a fault-bounded trough bordered on the south by rocks of the gneiss terrane (Fig. 8B).

Overlying the rocks of the pre-quartzite phase is a thick wedge of epiclastic rocks here called the quartzite phase. These rocks were derived from both the greenstone-granite terrane to the north and the gneiss terrane to the south of the basin (Peterman, 1966; Keighin and others, 1972; Morey, 1973). At the beginning of this phase, sedimentation was confined to a basin over the Great Lakes tectonic zone and the gneiss terrane (Fig. 8C), but later spilled over onto the greenstone-granite terrane to the north (Fig. 8D) as the width of the Animikie basin increased with time. Much of the quartzite phase is characterized by rather monotonous interbeds of quartz-rich wacke and arenite, siltstone, and mudstone, which are punctuated by scattered beds of sandy limestone, dolomite, and quartz-pebble grit or conglomerate. A generally increasing ratio of mud to sand toward the axis of the basin and stratigraphically upward implies that subsidence was greatest along the axis of the basin. However, all of these rocks appear to have shallow-water attributes which imply that sedimentation more or less kept pace with subsidence (Morey, 1973).

Rocks of the quartzite phase, as well as those of the greenstone-granite terrane itself, provided the foundation for a southward-facing shelf (Fig. 8E) on which were deposited various kinds of iron-formation having dominantly "shallow-water" attributes to the north and west and "deeper water" attributes to the south and east (Morey, 1973; 1983). These complexly intercalated facies have been attributed to deposition under stable tectonic conditions near a transgressing and regressing strandline (White,

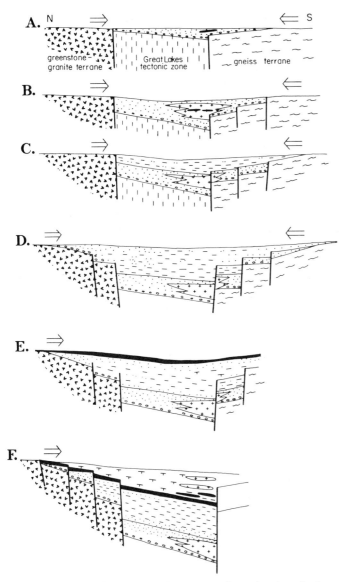

Figure 8. Schematic north-south cross sections showing the inferred evolution of the northwestern segment of the Animikie basin. A, Denham time; B, Glen Township time; C, Little Falls time; D, Mahnomen-Pokegama time; E, Trommald-Biwabik time; F, Rabbit Lake-Virginia-Thomson time. Arrows indicate dominant direction of sediment transport.

1954). However, the presence of what appear to be "growth faults" in the Biwabik Iron Formation (e.g., Gundersen and Schwartz, 1962, Pl. 1) attests to at least some tectonic instability during iron-formation deposition.

The shelf deposits are overlain by dominantly black, carbonaceous, laminated mudstone deposited in the starved-basin environment of the fourth or transitional phase. As the shelf continued to subside into deeper water (Fig. 8F), mud deposition was periodically interrupted by southward-flowing turbidity currents that deposited beds of graywacke and siltstone in a series of southward-prograding submarine fan complexes (Lucente and Morey, 1983).

The pivotal event in the depositional history of the basin was the foundering of the shelf. Prior to that event, deposition in the basin was more or less symmetrical with detritus being supplied to the basin from both the north and the south. However, after the shelf subsided during the transitional phase, the sedimentological record became markedly asymmetrical, and the rocks of the flysch phase constituted what appears to have been a southward-facing wedge (Fig. 8F) of considerable thickness.

Compressional Stage

Deformation terminating deposition of the stratified rocks was complex and probably the result of regional, principal stresses nearly perpendicular to the trend of the Great Lakes tectonic zone. The prevailing east-northeast-trends of folds and their associated axial-plane cleavages and foliations indicate a dominant direction of tectonic transport to the northwest.

Early metamorphism accompanied folding, as shown by a regional subhorizontal metamorphic schistosity developed early in the deformation of the Thomson Formation, prior to folding about generally east-trending axes (Connolly, 1981). Continued metamorphism through the main period of folding is indicated by garnet and staurolite isograds that on a regional scale appear to form annular patterns around domes cored by gneissic basement rocks metamorphosed to the lower amphibolite facies. This relationship, combined with the fact that cataclastic structural elements in the gneissic cores are subconformable with those in the stratified rocks, implies that the formation of the gneiss domes and the metamorphism of the fringing supracrustal rocks to the amphibolite facies were virtually contemporaneous.

Folding was accompanied or followed by dip-slip reverse faulting with south side up along the northern edge of the Great Lakes tectonic zone. The southern boundary fault of the Great Lakes tectonic zone also must be a south-side-up, high-angle reverse fault, inasmuch as an unknown, but presumably considerable amount of strata was eroded from this part of the basin after emplacement of the Penokean igneous rocks. However, the timing of these faulting events relative to the Penokean orogeny has not been documented.

All of the structural evidence suggests that deformation was accompanied by a considerable amount of crustal shortening, although the amount of shortening and its underlying causes are not known. The spatial coincidence of the Penokean foldbelt with the Great Lakes tectonic zone and the presence of gneiss domes just to the south of the zone may imply that at least some of the folding resulted from uplift and expansion of the gneissic basement rocks and concurrent lateral transport of the overlying supracrustal rocks against the more rigid greenstone-granite crust to the northwest.

Many of the plutonic intrusive rocks of early Proterozoic age in east-central Minnesota appear to have been emplaced during the waning stages of the Penokean orogeny. In particular, the gabbroic, dioritic, and granodioritic bodies contain incipient cataclastic structures that coincide with fold axes and other structural elements formed in the stratified rocks during the Penokean orogeny (Fig. 7). These rock bodies therefore are late syntectonic in age. In contrast, the somewhat younger granitic rocks are relatively homogeneous and undeformed. Furthermore, they form a composite pluton of batholithic dimensions that is elongate in a northeastward direction, an orientation somewhat at an angle to the Great Lakes tectonic zone and to the direction of extension expected from the decay of a regional northwest-southeast compressive stress. Because the pluton is at an oblique angle to the tectonic grain of the Penokean foldbelt, it may reflect a tectonic event somewhat younger than the Penokean orogeny that has not yet been dated by modern radiometric methods.

COMPARISON WITH THE SOUTHEASTERN SEGMENT OF THE ANIMIKIE BASIN

Many authors (for example Van Hise and Leith, 1911; Leith and others, 1935; James, 1960; Bayley and James, 1973; Sims, 1976; Morey, 1978) have emphasized the similarity of the lower Proterozoic stratigraphic succession in Minnesota to that of the Marquette Range Supergroup in northern Wisconsin and Upper Michigan (Fig. 3). There the Chocolay Group unconformably overlies Archean rocks and consists of a basal conglomerate overlain by quartzite and carbonate rocks. The Menominee Group, which unconformably overlies the Chocolay Group, contains quartzite, shale and mudstone, graywacke, volcanic rocks, and the major iron-formations of the region. The Baraga Group conformably (Wisconsin) to unconformably (Upper Michigan) overlies the Menominee Group. The dominant rock types of the Baraga Group are mudstone and shale, graywacke, iron-formation, and volcanic rocks. The Paint River Group in the southeastern part of Upper Michigan has no stratigraphic analog in Minnesota. It unconformably overlies the Baraga Group, and consists of graywacke, mudstone, iron-formation, and thick intercalations of volcanic rocks, and as such represents a resumption of the depositional regime characteristic of the Baraga Group.

Despite the broad stratigraphic similarities, there are some marked differences between the two segments of the Animikie basin. Sedimentation south of the Great Lakes tectonic zone in the southeastern segment occurred in a number of fault-bounded basins that are only tens of kilometers long, whereas sedimentation in the northwestern segment occurred within a single basin hundreds of kilometers long. LaRue and Sloss (1980) suggested that the strata in the northwestern segment were perhaps deposited in an aulacogen or in a wrench-fault basin, whereas the strata of the Marquette Range Supergroup in the southeastern segment can best be explained as sedimentation on, and deformation of, a rifted passive margin as proposed by Cambray (1977, 1978a, 1978b).

Geologic observations outlined in the preceding sections are not consistent with the contention that the northwestern segment of the Animikie basin formed as an aulacogen. Although the evolution of the northwestern segment involved some vertical crustal movements, other attributes typically associated with aulacogens (Hoffman, 1973; Hoffman and others, 1974; Burke, 1977), such as the longitudinal transport of sediments along the axis of the basin, the presence of subalkalic or alkalic volcanic rocks, regional transcurrent faults, or sediments that thicken toward a platform margin, are all lacking.

The northwestern and southeastern segments of the basin exhibit vastly different levels of erosion; therefore, the apparent difference in structural style may be more apparent than real. The tectonic attributes of the two segments (Fig. 9) differ chiefly in magnitude and intensity rather than in style. Therefore, it seems likely that sedimentation in both segments occurred within similar tectonic regimes and that any tectonic hypothesis to account for the evolution of one segment of the basin must also account for the evolution of the other segment. In addition, because sedimentation and deformation can be assumed to be a continuum within the same tectonic cycle, any hypothesis to account for extension in the basin must also account for the compression.

TECTONIC IMPLICATIONS

It now seems certain that the lower Proterozoic rocks in the Animikie basin were deposited in an enlarging intracontinental basin. The abrupt changes in sedimentological and tectonic patterns occurring over the boundary between contrasting kinds of Archean basement rocks imply that the northwestern and southeastern segments had vastly different tectonic stabilities in early Proterozoic time. The Great Lakes tectonic zone in Minnesota and Wisconsin appears to have been a zone of weakness in early Proterozoic time where crustal extension, faulting, and concurrent subsidence created a riftlike depression in which the sediments accumulated. In Michigan, sedimentation during the extensional stage occurred in linear, discontinuous, and diversely oriented basins developed over gneissic basement rocks. These basins appear to have resulted from the subsidence of discrete blocks bounded by preexisting basement structures (Cannon, 1973). During the compressional stage, the gneissic basement rocks in both segments were uplifted as gneiss

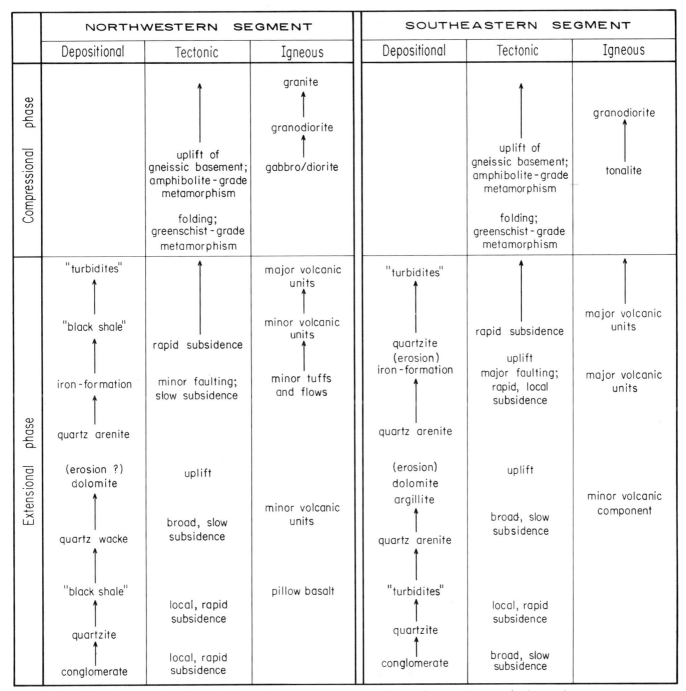

Figure 9. Temporal distribution of depositional, tectonic, and igneous elements in the north-western and southeastern segments of the Animikie basin. Arrows represent transitional events.

domes and fault-bounded blocks along the same basement structures. Nodal metamorphism was contemporaneous with uplift and was superimposed on a regional low-grade metamorphism developed during folding.

Although an internally consistent scenario can be proposed for the history of the entire Animikie basin, there is no consensus as to the driving forces that caused the basin to form or as to the cause of the deformational, metamor-

phic, and igneous events associated with the Penokean orogeny. Despite the variations imposed on the overall evolution of the basin by the contrasting kinds of basement rocks, the stratigraphic sequence strongly resembles that of Phanerozoic geosynclines, as first noted by Pettijohn (1957, p. 640–641) in his textbook *Sedimentary Rocks*. However, the significance of Pettijohn's observation was not appreciated until it was established that Phanerozoic geosynclines form by plate-tectonic processes. The similarity of the sedimentary fill in the Animikie basin to that in Phanerozoic geosynclines and the presence of extensive volcanic sequences south of the basin have led to several attempts to explain the evolution of the Animikie basin by plate-tectonic processes characteristic of Phanerozoic time. For example, Van Schmus (1976) hypothesized that a north-dipping subduction zone existed south of the Animikie basin in early Proterozoic time. In this hypothesis, volcanic rocks south of the basin represent the island-arc region, and the granitic rocks the eroded roots of that arc. The rocks of the basin itself would then occur between the island arc to the southeast and the shoreline to the northwest, the rocks in northern Minnesota being closest to the shoreline. The Penokean orogeny in this hypothesis was the product of a consuming continental margin with ocean floor to the south subducted toward the north under the Animikie basin.

It is now known that all of the lower Proterozoic rocks, including those of volcanic affinity in northern Wisconsin and east-central Minnesota, were deposited on continental crust (Cannon, 1973; Sims, 1976; Morey, 1978; Van Schmus and Anderson, 1977; Sims and Peterman, 1983). Furthermore Van Schmus and Anderson (1977) have shown that many of the gneissic rocks of central Wisconsin are Archean in age. Plutons of Penokean age, which intrude these Archean rocks (Van Schmus, 1980), indicate that this terrane was extensively involved in the Penokean orogeny. Thus the idea that oceanic crust existed to the south of the Animikie basin during early Proterozoic time is no longer tenable (Van Schmus and Bickford, 1981).

The fact that the stratified rocks were deposited on continental crust does not, however, preclude the possibility that the Animikie basin formed by rifting processes somewhat akin to those proposed to explain the opening of a protoceanic basin. In this hypothesis, rifting and abrupt subsidence were followed by the formation of an expanding oceanic plate, which in turn was followed by reversal of plate movement, subduction, compression, and metamorphism (Cambray, 1977, 1978a, 1978b). The Penokean orogeny would then reflect the ultimate closing of the basin by subduction of newly formed oceanic crust between two colliding continental plates.

An intracontinental rifting model in which the Animikie basin represents a gradually opening rift zone and the volcanic rocks of central Wisconsin represent newly formed oceanic crust has many appealing sedimentological aspects. However, it also has some problems. Sedimentological data from rocks correlative with the Animikie Group on the Gogebic range in northern Wisconsin indicate that a craton existed somewhere to the south of the Gogebic range (Alwin, 1976; Ojakangas, 1982) and that the axis of the Animikie basin was located between the Gogebic range and the Mesabi range in Minnesota. Furthermore, the volcanic

rocks of northeastern Wisconsin overlie sedimentary rocks of shallow-water affinity like those exposed to the north in the main part of the basin (Schulz and Sims, 1982). Thus there is no substantive evidence either that rifting proceeded to the stage where extensive volcanic sequences were developed between disrupted crustal segments or that the Penokean orogeny reflects the ultimate closing of the basin by subduction of those sequences between two colliding continental plates. Geologic phenomena indicative of a suture between two colliding plates (e.g., Moores, 1981), such as large foreland-directed overthrusts and associated melanges, paired high-pressure/low-temperature and low-pressure/high-temperature metamorphic belts, ophiolites, or sheeted gabbroic complexes have not been documented, despite a considerable amount of modern geologic mapping (e.g., Morey and others, 1982, and contained references).

CONCLUSIONS

The geologic evidence implies that the lower Proterozoic rocks of the Lake Superior region were deposited and deformed in an intracontinental setting along a tectonically active rift zone that involved extension between two contrasting plates and formation of a basin and geosynclinal deposits like those formed in Phanerozoic time. However, the geologic evidence also implies that the tectonic processes during the compressional stage were somewhat unlike those associated with collision-type orogenies.

I suggest that the Animikie basin has all of the major components of an arrested protoceanic rift system that did not reach the stage where new oceanic crust formed or where subduction of such crust occurred. This hypothesis differs from those of Cambray (1977, 1978a, 1978b) only as to the extent to which the Animikie basin opened and consequently as to the origin of the thick volcanic sequences south of the basin. Because volcanism occurred both early and late in the history of the Animikie basin, these volcanic sequences could have formed either early in the extensional stage or later in the history of the main part of the basin as extension progressed southward. They need not, in either case, represent the remains of new oceanic crust consumed by subduction.

The fact that the overall history of the Animikie basin has some attributes that set it apart from a "typical" Phanerozoic geosyncline does not negate the usefulness of plate-tectonic processes in explaining the evolution of the basin. Differences among Phanerozoic geosynclines imply that cause-effect relationships between deep-seated tectonic forces and the near-surface manifestations of those forces are not invariably the same, either in space or in time. Thus each geosyncline can be considered an entity formed by a specific combination of tectonic processes that may never have been repeated in exactly the same way. Moreover it still is not clear as to what driving forces caused the plate interactions we see, either in Phanerozoic geosynclines or in the Animikie basin. Thus we still do not know if the evolution of the Animikie basin followed the same "rules" of extension and compression that govern the evolution of Phanerozoic geosynclines or if the basin formed by a sequence of processes unique to the Lake Superior region in early Proterozoic time. Resolving this fundamental problem will require new knowledge gathered in detail from the rocks themselves.

ACKNOWLEDGMENTS

The ideas summarized in this paper are the results of discussions with numerous colleagues over the past 10 years. Many of the people are referenced in the text, but I am particularly indebted to R. W. Ojakangas, P. K. Sims, D. L. Southwick, and Matt Walton for stimulating and critical comment. This particular report has benefited greatly from the typing skills of Linda McDonald, the editorial comments of N. Balaban, the drafting of Richard Darling, and the critical comments of C. Craddock and T. B. Holst.

REFERENCES CITED

Aldrich, L. T., Davis, G. L., and James, H. L., 1965, Ages of minerals from metamorphic and igneous rocks near Iron Mountain, Michigan: Journal of Petrology, v. 6, p. 445–472.

Alwin, B. W., 1976, Sedimentation of the middle Precambrian Tyler Formation of north-central Wisconsin and northwestern Wisconsin [M.S. thesis]: Duluth, University of Minnesota, 170 p.

Bayley, R. W., and James, H. L., 1973, Precambrian iron-formations of the United States: Economic Geology, v. 68, p. 934–959.

Bayley, R. W., Dutton, C. E., and Lamey, C. A., 1966, Geology of the Menominee iron-bearing district, Dickinson County, Michigan, and Florence and Marinette Counties, Wisconsin: U.S. Geological Survey Professional Paper 513, 96 p.

Beck, W., and Murthy, V. R., 1982, Rb-Sr and Sm-Nd isotopic studies of Proterozoic mafic dikes in northeastern Minnesota [abs.]: Institute on Lake Superior Geology, 28th Annual, International Falls, Minnesota, Proceedings, p. 5.

Burke, K., 1977, Aulacogens and continental breakup: Annual Review of Earth and Planetary Sciences, v. 5, p. 371–396.

Cambray, F. W., 1977, The geology of the Marquette district—a field guide: Michigan Basin Geological Society, 62 p.

—— 1978a, Plate tectonics as a model for the environment of sedimentation of the Marquette Range Supergroup and the subsequent deformation and metamorphism associated with the Penokean orogeny [abs.]: Institute on Lake Superior Geology, 24th Annual, Milwaukee, Wisconsin, Abstracts and Field Guides, p. 6.

—— 1978b, Plate tectonics as a model for the development of deposition and deformation of the early Proterozoic (Precambrian X) of northern Michigan: Geological Society of America Abstracts with Programs, v. 10, p. 376.

Cannon, W. F., 1973, The Penokean orogeny in northern Michigan, *in* Young, G. M., ed., Huronian stratigraphy and sedimentation: Geological Association of Canada Special Paper 12, p. 251–271.

Cannon, W. F., and Gair, J. E., 1970, A revision of stratigraphic nomenclature for middle Precambrian rocks in northern Michigan: Geological Society of America Bulletin, v. 81, p. 2843–2846.

Chandler, V. W., Watts, A. B., and Gulbranson, B. L., 1982, Magnetic anomaly studies of the northern Animikie basin [abs.]: Institute on Lake Superior Geology, 28th Annual, International Falls, Minnesota, Proceedings, p. 7.

Connolly, M. R., 1981, The geology of the middle Precambrian Thomson Formation in southern Carlton County, Minnesota [M.S. thesis]: Duluth, University of Minnesota, 133 p.

Dott, R. J., Jr., and Dalziel, I.W.D., 1972, Age and correlation of the Precambrian Baraboo Quartzite of Wisconsin: Journal of Geology, v. 80, p. 552–568.

Floran, R. J., and Papike, J. J., 1975, Petrology of the low-grade rocks of the Gunflint Iron-formation, Ontario-Minnesota: Geological Society of America Bulletin, v. 86, p. 1169–1190.

French, B. M., 1968, Progressive contact metamorphism of the Biwabik Iron-formation, Mesabi Range, Minnesota: Minnesota Geological Survey Bulletin 45, 103 p.

Goldich, S. S., 1968, Geochronology of the Lake Superior region: Canadian Journal of Earth Sciences, v. 5, p. 715–724.

Goldich, S. S., Nier, A. O., Baadsgaard, H., Hoffman, J. H., and Krueger, H. W., 1961, The Precambrian geology and geochronology of Minnesota: Minnesota Geological Survey Bulletin 41, 193 p.

Gundersen, J. N., and Schwartz, G. M., 1962, The geology of the metamorphosed Biwabik Iron-formation, eastern Mesabi district, Minnesota: Minnesota Geological Survey Bulletin 43, 139 p.

Han, Tsu-ming, 1968, Ore mineral relations in the Cuyuna sulfide deposit, Minnesota: Mineralium Deposita [Berlin], v. 3, p. 109–134.

Hoffman, P., 1973, Evolution of an early Proterozoic continental margin: The Coronation geosyncline and associated aulacogens of the northwestern Canadian Shield: Royal Society of London, Philosophical Transactions, ser. A, v. 273, p. 547–581.

Hoffman, P., Dewey, J. F., and Burke, K., 1974, Aulacogens and their genetic relation to geosynclines with a Proterozoic example for Great Slave Lake, Canada, *in* Dott, R. H., and Shaver, R. H., eds., Modern and ancient geosynclinal sedimentation: Society of Economic Paleontologists and Mineralogists Special Publication 19, p. 38–55.

James, H. L., 1954, Sedimentary facies of the Lake Superior iron-formations: Economic Geology, v. 49, p. 253–293.

—— 1955, Zones of regional metamorphism in the Precambrian of northern Michigan: Geological Society of America Bulletin, v. 66, p. 1455–1488.

—— 1958, Stratigraphy of pre-Keweenawan rocks in parts of northern Michigan: U.S. Geological Survey Professional Paper 314-C, p. C27–C44.

—— 1960, Problems of stratigraphy and correlation of Precambrian rocks with particular reference to the Lake Superior region: American Journal of Science (Bradley volume), v. 258-A, p. 104–114.

Keighin, C. W., Morey, G. B., and Goldich, S. S., 1972, East-central Minnesota, *in* Sims, P. K., and Morey, G. B., eds., Geology of Minnesota: A centennial volume: Minnesota Geological Survey, p. 240–255.

King, P. B., 1969, Tectonic map of North America: U.S. Geological Survey, scale 1:5,000,000.

LaRue, D. K., and Sloss, L. L., 1980, Early Proterozoic sedimentary basins of the Lake Superior region: Geological Society of America Bulletin, Part. I, v. 91, p. 450–452.

Leith, C. K., Lund, R. J., and Leith, A., 1935, Pre-Cambrian rocks of the Lake Superior region: U.S. Geological Survey Professional Paper 184, 34 p.

Lucente, M. E., and Morey, G. B., 1983, Stratigraphy and sedimentology of the lower Proterozoic Virginia Formation (Animikie Group), northern Minnesota: Minnesota Geological Survey Report of Investigations 28, 28 p.

Marsden, R. W., 1972, Cuyuna district, *in* Sims, P. K., and Morey, G. B., eds., Geology of Minnesota: A centennial volume: Minnesota Geological Survey, p. 227–239.

McSwiggen, P. L., 1982, Uranium in lower Proterozoic phosphate-rich metasedimentary rocks of east-central Minnesota [abs.]: Institute on Lake Superior Geology, 28th Annual, International Falls, Minnesota, Proceedings, p. 25.

Moores, E. M., 1981, Ancient suture zones within continents: Science, v. 213, no. 4503, p. 41–46.

Morey, G. B., 1969, The geology of the middle Precambrian Rove Formation in northeastern Minnesota: Minnesota Geological Survey Special Publication SP-7, 62 p.

——1973, Stratigraphic framework of middle Precambrian rocks in Minnesota, in Young, G. M., ed., Huronian stratigraphy and sedimentation: Geological Association of Canada Special Paper 12, p. 211–249.

——1978, Lower and middle Precambrian stratigraphic nomenclature for east-central Minnesota: Minnesota Geological Survey Report of Investigations 21, 52 p.

——1979, Stratigraphic and tectonic history of east-central Minnesota, in Balaban, N. H., ed., Field trip guidebook for stratigraphy, structure and mineral resources of east-central Minnesota: Minnesota Geological Survey Guidebook Series 9, p. 13–28.

——1983, Animikie basin, Lake Superior region, USA, in Trendall, A. F., ed., Iron-formations: Facts and problems: Elsevier (in press).

Morey, G. B., and Ojakangas, R. W., 1970, Sedimentology of the middle Precambrian Thomson Formation, east-central Minnesota: Minnesota Geological Survey Report of Investigations 13, 32 p.

Morey, G. B., and Sims, P. K., 1976, Boundary between two Precambrian W terranes in Minnesota and its geologic significance: Geological Society of America Bulletin, v. 87, p. 141–152.

Morey, G. B., Olsen, B. M., and Southwick, D. L., 1981, Geologic map of Minnesota, east-central Minnesota: Minnesota Geological Survey, scale 1:250,000.

Morey, G. B., Sims, P. K., Cannon, W. F., Mudrey, M. G., Jr., and Southwick, D. L., 1982, Geologic map of the Lake Superior region, bedrock geology: Minnesota Geological Survey State Map Series S-13, scale 1:1,000,000.

Ojakangas, R. W., 1982, Tidal deposits in the early Proterozoic basin of the Lake Superior region—the Palms and Pokegama Formations: Evidence for subtidal shelf deposition of Superior-type banded iron-formation [abs.]: Institute on Lake Superior Geology, 28th Annual, International Falls, Minnesota, Proceedings, p. 34.

Perry, E. C., Jr., Tan, F. C., and Morey, G. B., 1973, Geology and stable isotope geochemistry of the Biwabik Iron Formation, northern Minnesota: Economic Geology, v. 68, p. 1110–1125.

Peterman, Z. E., 1966, Rb-Sr dating of middle Precambrian metasedimentary rocks of Minnesota: Geological Society of America Bulletin, v. 77, p. 1031–1041.

——1979, Geochronology and the Archean of the United States: Economic Geology, v. 74, p. 1544–1562.

Pettijohn, F. J., 1957, Sedimentary rocks: Harper and Row, 718 p.

Schmidt, R. G., 1958, Titaniferous sedimentary rocks in the Cuyuna district, central Minnesota: Economic Geology, v. 53, p. 708–721.

——1959, Bedrock geology of the northern and eastern parts of the North range, Cuyuna district, Minnesota: U.S. Geological Survey Mineral Investigations Field Studies Map MF-182, scale 1:7,200.

——1963, Geology and ore deposits of Cuyuna North range, Minnesota: U.S. Geological Survey Professional Paper 407, 96 p.

Schulz, K. J., and Sims, P. K., 1982, Nature and significance of shallow water sedimentary rocks in northeastern Wisconsin [abs.]: Institute on Lake Superior Geology, 28th Annual, International Falls, Minnesota, Proceedings, p. 43.

Shoemaker, E. M., ed., 1975, Continental drilling: Report of the workshop on continental drilling, Ghost Ranch, Abiquiu, New Mexico, June 10–13, 1974: Carnegie Institution of Washington, 56 p.

Sims, P. K., 1976, Precambrian tectonics and mineral deposits, Lake Superior region: Economic Geology, v. 71, p. 1092–1127.

——1980, Boundary between Archean greenstone and gneiss terranes in northern Wisconsin and Michigan, in Morey, G. B., and Hanson, G. N., eds., Selected studies of Archean gneisses and lower Proterozoic rocks, southern Canadian Shield: Geological Society of America Special Paper 182, p. 113–124.

Sims, P. K., and Morey, G. B., 1973, A geologic model for the development of early Precambrian crust in Minnesota [abs.]: Geological Society of America Abstracts with Programs, v. 5, p. 812.

Sims, P. K., and Peterman, Z. E., 1980, Geology and Rb-Sr age of lower Proterozoic granitic rocks in northern Wisconsin, in Morey, G. B., and Hanson, G. N., eds., Selected studies of Archean gneisses and lower Proterozoic rocks, southern Canadian Shield: Geological Society of America Special Paper 182, p. 139–146.

——1983, Evolution of the Penokean foldbelt, Lake Superior region, and its tectonic environment, in Medaris, L. G., Jr., ed., Early Proterozoic Geology of the Great Lakes Region: Geological Society of America Memoir 160 (this volume).

Sims, P. K., Card, K. D., Morey, G. B., and Peterman, Z. E., 1980, The Great Lakes tectonic zone—a major crustal structure in central North America: Geological Society of America Bulletin, Part I, v. 91, p. 690–698.

Sims, P. K., Card, K. D., and Lumbers, S. B., 1981, Evolution of early Proterozoic basins of the Great Lakes region, in Campbell, F.A.H., ed., Proterozoic basins of Canada: Geological Survey of Canada Paper 81-10, p. 379–397.

Smith, E. I., 1978, Precambrian rhyolites and granites in south-central Wisconsin, field relations and geochemistry: Geological Society of America Bulletin, v. 89, p. 875–890.

Stockwell, C. H., McGlynn, J. C., Emslie, R. F., Sanford, B. V., Norris, A. W., Donaldson, J. A., Fahrig, W. F., and Currie, K., 1970, Geology of the Canadian Shield, in Douglas, R.J.W., ed., Geology and economic minerals of Canada: Geological Survey of Canada, Economic Geology Report 1, p. 44–150.

Trendall, A. F., 1968, Three great basins of Precambrian banded iron formation deposition: A systematic comparison: Geological Society of America Bulletin, v. 79, p. 1527–1544.

Van Hise, C. R., and Leith, C. K., 1911, The geology of the Lake Superior region: U.S. Geological Survey Monograph 52, 641 p.

Van Schmus, W. R., 1976, Early and middle Proterozoic history of the Great Lakes area, North America: Royal Society of London, Philosophical Transactions, ser. A, v. 280, p. 605–628.

——1978, Geochronology of the southern Wisconsin rhyolites and granites: Wisconsin Geological and Natural History Survey, Geoscience Wisconsin, v. 2, p. 19–24.

——1980, Chronology of igneous rocks associated with the Penokean orogeny in Wisconsin, in Morey, G. B., and Hanson, G. N., eds., Selected studies of Archean gneisses and lower Proterozoic rocks, southern Canadian Shield: Geological Society of America Special Paper 182, p. 159–168.

Van Schmus, W. R., and Anderson, J. L., 1977, Gneiss and migmatite of Archean age in the Precambrian basement of central Wisconsin: Geology, v. 5, p. 45–48.

Van Schmus, W. R., and Bickford, M. E., 1981, Proterozoic chronology and evolution of the Midcontinent region, North America, in Kroner, A., ed., Precambrian plate tectonics: Elsevier, p. 261–296.

Van Schmus, W. R., Medaris, L. G., Jr., and Banks, P. O., 1975, Geology and age of Wolf River batholith, Wisconsin: Geological Society of America Bulletin, v. 86, p. 907–914.

Weiblen, P. W., 1964, A preliminary study of the metamorphism of the Thomson Formation [M.S. thesis]: Minneapolis, University of Minnesota, 153 p.

White, D. A., 1954, Stratigraphy and structure of the Mesabi range, Minnesota: Minnesota Geological Survey Bulletin 38, 92 p.

MANUSCRIPT ACCEPTED BY THE SOCIETY MARCH 4, 1983

Geological Society of America
Memoir 160
1983

Geochemistry and evolution of the early Proterozoic, post-Penokean rhyolites, granites, and related rocks of south-central Wisconsin, U.S.A.

Eugene I. Smith
Department of Geoscience
University of Nevada, Las Vegas
Las Vegas, Nevada 89154

ABSTRACT

Four chemically and mineralogically distinct rock suites formed in south-central Wisconsin during a major anorogenic intrusive-extrusive event 1.76 b.y. ago. This anorogenic event followed by at least 60 m.y. the terminal calc-alkalic plutonism and volcanism of the Penokean orogeny. The 1.76-b.y.-old rocks may represent the only surface expression in the Lake Superior area of a 1.69 to 1.78-b.y.-old felsic terrain that can be traced in the subsurface from Wisconsin into northern Illinois and possibly as far as western Arizona. The magma types formed during the anorogenic event are (1) a peraluminous suite of texturally variable ash-flow tuffs; (2) a metaluminous suite containing quartz- and orthoclase-bearing rhyolites and closely associated granophyric granites; (3) a suite of low SiO_2, high Sr, and REE depleted granites at Baxter Hollow and chemically similar rhyolite dikes on Observatory Hill and at Marquette (Baxter Hollow suite); and (4) a suite containing a diorite intrusion and numerous dikes of tholeiitic basalt and andesite (Denzer suite).

Major and trace element modeling studies suggest the following scenario for the formation of these post-Penokean rocks. The peraluminous and metaluminous suites were probably derived by partial melting (16%) of different crustal sources of tonalitic composition having slightly different residue mineralogies. These source rocks may have formed earlier by igneous activity related to the Penokean orogeny. Compositional variation within the two suites may be related to feldspar-dominated fractional crystallization. The Baxter Hollow suite was probably generated by large-scale (66%) partial melting of a crustal source of dioritic composition. Heat for such a major melting episode may have been provided by magma of the Denzer suite rising from a mantle source. Modeling studies suggest that 10% melting of either garnet peridotite, or eclogite will provide liquid of the composition of the Denzer suite. Fractional crystallization is most probably responsible for the chemical variation observed within both the Denzer and Baxter Hollow suites.

INTRODUCTION

Four chemically and mineralogically distinct rock suites formed in south-central Wisconsin during a major post-Penokean anorogenic-intrusive event (1.76 b.y. ago; Van Schmus, 1978). These are (1) a peraluminous suite of texturally variable ash-flow tuffs; (2) a metaluminous suite containing quartz- and orthoclase-bearing rhyolites and granophyric granites; (3) a suite of low SiO_2, high Sr, and REE depleted granites at Baxter Hollow and chemically similar rhyolite dikes on Observatory Hill and at Marquette (the Baxter Hollow suite); and (4) a suite consisting

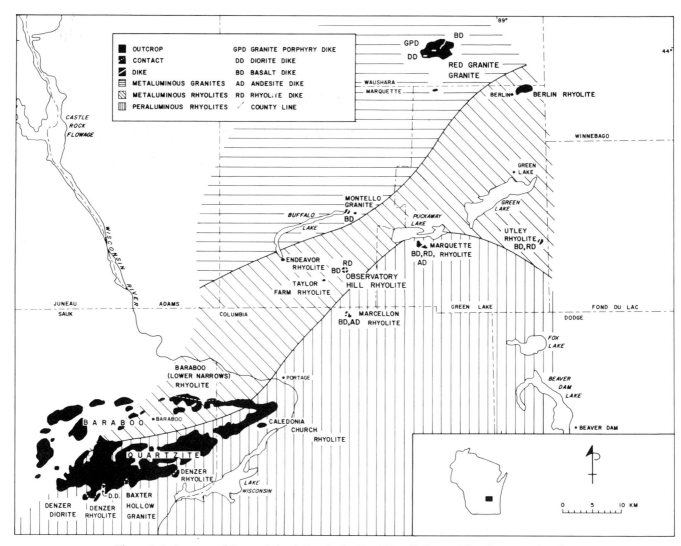

Figure 1. Index map of Precambrian inliers in the Fox River Valley and Baraboo area showing the distribution of the four rock suites. Since contacts between rock suites roughly parallel structures in exposures, patterns on this map probably reflect the geology of the buried Precambrian basement in this area.

of a diorite intrusion and numerous dikes varying in composition from tholeiitic basalt to andesite (the Denzer suite) (Fig. 1). Rocks of the four suites may represent the only surface expresssion in the Lake Superior area of a 1.69- to 1.78-b.y.-old felsic terrain that can be traced in the subsurface from Wisconsin into northern Illinois. Rocks of a similar age extend in the subsurface and in outcrop as far as western Arizona (Van Schmus and Bickford, 1981). In Wisconsin, exposures of rhyolite and granite occur as inliers that are surrounded and partially covered by Cambrian sandstone and Pleistocene drift and outwash. The basement terrain between exposures has been sampled in many places by deep water wells. Petrographic study of cuttings from these wells demonstrates that granites and rhyolites, lithologically similar to those in the inliers, occur in the subsurface between inliers (Smith, 1978b). Lithological,

geochronological, and geochemical data collectively suggest that all of the inliers are part of the same composite pluton.

The rhyolites of both the peraluminous and metaluminous suites occur mainly as ash-flow tuffs and characteristically display eutaxitic texture. Ash-flow cooling units are commonly interbedded with volcaniclastic sedimentary rocks that include sections of debris-flow breccia. The associated granites display granophyric texture and may have intruded their own volcanic cover. The granites occur mainly to the north of the rhyolite exposures. Anderson and others (1980) presented a chemical study of the granites and recognized metaluminous and peraluminous suites for these south-central Wisconsin plutonic rocks. Also included in their study are four granite plutons in northern Wisconsin that are coeval with the south-central Wisconsin granites (Van Schmus, 1980). The northern Wisconsin granites may represent the deep-seated equivalents of those in southern Wisconsin (Van Schmus and Bickford, 1981). The rhyolite ash-flow tuffs described here and in Smith (1978a) belong to the same magma types proposed by Anderson and others (1980).

Although relatively unmetamorphosed, the Rb/Sr systems of the 1.76-b.y.-old rhyolites have been reset at 1.63 b.y. (Van Schmus and others, 1975). This event probably represents a thermal event that correlated with the folding of the volcanic rocks (Smith, 1978a) and the overlying sedimentary pile (Dalziel and Dott, 1970; Brandon and others, 1980). In general, metamorphism did not reach higher than greenschist facies; however, local igneous activity resulted in amphibolite grade metamorphism of the Waterloo Quartzite (Geiger and others, 1982; Brandon and others, 1980). The 1.63-b.y.-old event may be a "foreland manifestation" of an extensive orogenic belt that occurs to the south (Van Schmus and Bickford, 1981).

The 1.76-b.y.-old event followed the extensive basin sedimentation and terminal plutonism, volcanism, and metamorphism of the Penokean orogeny (Goldich and others, 1961; Sims, 1976; Van Schmus, 1980) by at least 60 m.y. Igneous activity related to the Penokean orogeny peaked between 1.82 and 1.86 b.y. ago (Van Schmus, 1976, 1980) and is characterized by initial basaltic volcanism (Banks and Rebello, 1969) and subsequent calc-alkalic tonalite to adamellite magmatism (Cudzilo, 1978; Anderson and others, 1980). When the 1.76-b.y.-old peraluminous and metaluminous rhyolites and granites are compared to the igneous rocks of the earlier Penokean orogeny there are several important differences. The post-Penokean rocks are true granites and rhyolites and are too low in Ca, Mg, and Al and too rich in Fe and total alkalies, especially K, to be calc-alkalic. The four 1.76-b.y.-old suites may have formed during regional extension with conditions similar to the subsequent formation of the Wolf River Batholith at 1.5 b.y. (Anderson and Cullers, 1978) and the Keweenawan Rift at 1.1 b.y. (Green, 1977). Alternatively, they might have erupted during crustal relaxation after plate collision subsequent to the main phase of the Penokean orogeny (Petro, 1980).

The 1.76-b.y.-old rhyolites, granites, and related rocks of south-central Wisconsin record a critical change from the calc-alkalic activity of the Penokean event to anorogenic volcanism and plutonism that characterized post-Penokean time in the Lake Superior area. This paper describes the chemical characteristics of the four suites and presents models for their evolution. Model studies using rare earth elements (REE) and major elements suggest a scenario for the formation of the rhyolites and granites involving partial melting of a heterogeneous crust of intermediate composition with perhaps some melting of the mantle. The undifferentiated part of the Denzer suite may have been entirely derived by partial melting of mantle material.

ANALYTICAL METHODS

Major element and trace element data were published earlier by Smith (1978a and 1978b). The samples used for the REE determinations are from the same splits as the previous analyses; 43 samples were analyzed.

The REE data reported here were obtained by instrumental neutron activation analysis at the Phoenix Memorial Laboratory of the University of Michigan. The multi-element standards G-2 and AGV-1 were used as internal standards. Accuracy for REE with concentrations less than 1,000 ppm is 10%, except for Nd, Tb, and Yb, which is 20%.

Modal analysis was done by point counting of thin sections and in some cases by counting stained slabs. A minimum of 500 points were counted for each analyzed rock. A summary of the modal analyses is presented in Table 1. Rock names are from the chemical classification of Irvine and Baragar (1971) and were determined by a Fortran IV program by Smith and Stupak (1978).

Radiometric dates reported here are the U-Pb determinations of Van Schmus (1978, 1980) and Van Schmus and others (1975).

PETROGRAPHY AND SETTING

The rhyolites of the metaluminous suite are commonly plagioclase free and contain anhedral grains of quartz, and perthitic alkali feldspar. Grains of iron oxide, chlorite, clinozoisite, and a few euhedral zircon crystals are scattered randomly throughout the matrix. The matrix is coarsely devitrified to quartz and alkali feldspar, and as a result, no primary textures are preserved, although in several samples lenses of more coarsely devitrified matrix may represent flattened pumice fragments.

Peraluminous rhyolites belong to two geologically associated varieties. The first rhyolite type contains plagioclase (andesine) as the dominant phenocryst. The matrix is commonly devitrified with quartz and alkali feldspar as important constituents. Shards, broken spherulites, and a faint eutaxitic texture are the only primary textures preserved. Chlorite and sausserite are scattered throughout the matrix. The second rhyolite type contains large (up to 2 mm), rounded, and embayed grains of quartz associated with subhedral perthitic alkali feldspar. Plagioclase (andesine) is usually dusted with sausserite and occurs occasionally in glomeroporphyritic clots. The matrix is finely to coarsely devitrified, and few primary textures are preserved. Grains of iron oxide, chlorite, and apatite dot the matrix. Stratigraphically, the plagioclase-bearing rhyolites commonly alternate with the quartz-two feldspar units. This

TABLE 1. MODAL COMPOSITIONS OF THE SOUTH-CENTRAL WISCONSIN GRANITES, RHYOLITES AND RELATED ROCKS

Mineral	Peraluminous Suite				Metaluminous Suite				
	Marcellon Rhyolite* (2)	Marcellon Rhyolite (2)	Marquette Rhyolite (2)	Marquette Rhyolite (2)	Flynn's Quarry Dike (1)	Spring Lake Granite† (1)	Redgranite Granite (1)	Baxter Hollow Granite (1)	Denzer Diorite (1)
Quartz	6.0	–	9.4	–	9.6	27.9	21.9	20.7	10.8
Plagioclase	1.0	13.9	5.3	13.8	1.6	13.5	2.5	10.5	56.9
Orthoclase	4.0	–	13.6	–	19.2	57.1	70.7	61.4	4.4
Opaque	<1.0	<0.1	1.5	<0.1	<0.1	<0.1	0.5	<0.1	0.6
Biotite	–	–	–	–	6.3	0.7	3.4	7.1	–
Chlorite	–	–	–	–	2.6	0.2	0.2	–	10.5
Hornblende	–	–	–	–	1.0	–	–	–	13.3
Epidote	–	<0.1	1.3	<0.1	–	0.1	–	–	–
Fluorite	–	–	–	–	–	0.7	0.7	–	–
Apatite	–	–	–	–	–	<0.1	–	0.18	2.8
Groundmass	88.0	86.1	68.9	86.1	59.6	0.1	–	–	–
TOTAL	100.	100.	100.	100.	100.	100.	100.	100.	100.

Note: Modal analyses by point counting thin sections and stained slabs (500 points minimum per sample).

*Average of number of samples indicated.

†From Anderson and others (1980). Includes 0.1% zircon.

cyclic variation reflects either eruption from a differentiating source or compositional zonation within individual ash-flow cooling units (Smith, 1978a).

The Baxter Hollow granite (Gates, 1942) is exposed only beneath the cliffs of Baraboo Quartzite on the south limb of the Baraboo syncline; however, its stratigraphic relationship with respect to the quartzite is uncertain (Dalziel and Dott, 1970). The granite forms a small- to medium-sized stock that extends to the south in the subsurface for at least 20 km (Smith, 1978b). The granite is texturally and mineralogically variable (Gates, 1942), so that the rock used for this chemical study may not be representative of the stock as a whole. The granite contains intergrown grains of alkali feldspar and quartz, plagioclase, biotite, chlorite, and accessory zircon and apatite. The chemically similar rhyolite dike at Observatory Hill intrudes both the Observatory Hill (metaluminous) rhyolite and an east-trending basalt dike. This rhyolite is coarsely crystalline and contains plagioclase; quartz and alkali feldspar intergrown in a micropegmatitic texture; and clots of chlorite, epidote, and clinozoisite set in a fine-grained devitrified matrix (Table 1). It is separated from the Observatory Hill rhyolite by a 0.5-m-wide chill zone. The rhyolite dike at Marquette is composed of plagioclase phenocrysts (andesine) set in a fine-grained devitrified matrix.

The Denzer diorite (Stark, 1932) crops out in several isolated knobs to the south of the quartzite ridge forming the south range of the Baraboo syncline. It is composed of quartz, plagioclase altered to sausserite, hornblende, biotite altered to chlorite, perthitic alkali feldspar, iron oxide, and abundant subhedral to euhedral apatite. The high quartz

(10.8%), biotite (10.5%), and alkali feldspar (4.4%) content (Table 1) suggest that mineralogically the rock is a quartz diorite or granodiorite; however, chemically (Table 2) it appears to be a true diorite. The andesite and basalt dikes included within the Denzer suite intrude all rock types except the Observatory Hill rhyolite dike and the Baxter Hollow granite. The dikes are generally fine grained and contain plagioclase, epidote, and clinozoisite as common minerals. Coarse-grained varieties cross the metaluminous granite at Redgranite and the peraluminous rhyolite at Marcellon (Smith, 1978a).

Stratigraphic relationships indicate that the rocks of the Denzer and Baxter Hollow suites are slightly younger than the rhyolites and granites of the peraluminous and metaluminous suites. However, dating of the Observatory Hill rhyolite dike indicated no significant age difference between the dike and the 1.76-b.y.-old igneous rocks (Van Schmus, personal commun.).

CHEMICAL CHARACTERISTICS OF THE SUITES

Previous Work

Smith (1978a), in a study of major and trace elements, demonstrated that the post-Penokean rhyolites and granites belong to at least four chemical groups (Fig. 2). The Baxter Hollow granite stock and the coarse-grained rhyolite dikes on Observatory Hill form group 1. Group 2 consists of fine-grained and porphyritic rhyolites at the Marquette locality and fine-grained plagioclase-bearing rhyolites at the Marcellon exposure. Group 3 is composed

TABLE 2. SELECTED CHEMICAL ANALYSES FOR THE FOUR ROCK SUITES

	Peraluminous Suite			Metaluminous Suite						Baxter Hollow Suite			Denzer Suite		
	1	2	3	4	5	6	7	8	9	10	11	12	13	14	15
Major Oxides (%)															
SiO_2	71.93	71.80	75.98	75.83	76.14	75.09	75.60	72.68	72.10	70.26	69.28	69.33	56.21	48.94	60.59
TiO_2	0.25	0.27	0.13	0.21	0.24	0.21	0.17	0.25	0.30	0.30	0.42	0.43	1.33	0.99	0.93
Al_2O_3	14.38	14.23	11.89	11.92	11.79	12.33	12.59	12.15	12.74	14.10	13.90	14.23	13.20	17.84	16.47
Fe_2O_3	1.57	1.47	0.81	1.08	1.10	1.08	0.99		1.06	0.86	1.37	1.20	1.95	2.21	1.61
FeO	0.70	0.98	1.32	0.98	0.88	1.12	0.71	2.33	2.21	2.66	2.89	2.73	6.66	6.56	4.50
MnO	0.12	0.10	0.12	0.07	0.02	0.08	0.03	0.05	0.12	0.10	0.14	0.18	0.14	0.19	0.18
MgO	0.21	0.29	0.22	0.06	0.09	0.04	0.04	0.11	0.09	1.02	0.90		5.10	6.57	1.77
CaO	1.33	1.23	0.53	0.53	0.45	0.39	0.13	0.76	1.08	1.48	1.68	1.66	6.01	9.59	3.86
Na_2O	3.70	3.98	3.31	3.43	3.16	3.76	4.46	3.55	3.12	4.29	4.17	4.63	3.09	3.25	4.27
K_2O	4.44	4.30	4.69	5.45	5.65	5.58	4.95	5.59	6.12	3.69	4.33	3.41	2.56	1.11	3.27
H_2O^+	1.36	1.12	0.79	0.54	0.58	0.49	0.47	-	0.73	1.32	0.92	1.09	2.34	2.97	1.50
H_2O^-	0.06	0.06	0.04	0.01	0.01	0.00	0.02	-	0.03	0.04	0.07	0.05	0.14	0.12	0.11
P_2O_5	0.06	0.09	0.07	0.01	0.01	0.01	0.00	-	0.04	0.17	0.18	0.11	0.98	0.31	0.48
TOTAL	100.11	99.92	99.90	100.12	100.12	100.18	100.16	97.47	99.74	100.29	100.25	99.53	99.71	100.29	99.54
Elements															
Ba	710	661	246	440	390	410	545	947	1170	1200	1300	1500	1100	660	1200
Cr	13	18	52	38	8	23	16	-	26	25	32	29	150	42	16
Zr	207	224	186	420	590	590	480	-	590	320	320	290	620	75	220
Rb	153	125	128	190	202	152	120	214	180	120	165	98	75	38	105
Sr	131	126	64	25	21	31	28	24	56	324	210	212	625	642	514
La	51.6	47.3	49.3	82.0	98.5	65.6	62.0	76.6	68.9	38.2	55.7	46.7	40.0	13.3	40.3
Ce	113.4	90.0	115.3	200.4	225.2	169.4	146.2	161	157.6	79.3	125.6	103.6	92.3	29.4	89.0
Nd	37.5	37.8	45.3	65.1	74.5	58.9	53.7	-	59.2	22.6	50.8	37.0	41.4	11.9	45.4
Sm	7.0	6.7	9.0	11.4	13.8	11.3	10.3	13.4	10.4	4.3	6.7	6.7	7.3	0.2	7.1
Eu	0.9	1.8	0.4	0.9	0.7	1.1	0.9	1.37	2.1	1.2	1.9	1.9	2.3	1.6	2.3
Tb	1.1	0.9	1.8	2.4	2.7	2.3	1.8	2.31	2.2	0.6	0.9	0.9	0.6	0.5	1.2
Yb	3.4	3.6	4.9	11.2	9.1	9.8	7.8	7.8	6.1	1.3	2.5	2.5	1.3	0.9	1.4
Lu	0.6	0.5	0.9	1.1	1.4	0.9	0.8	1.0	1.0	0.3	0.5	0.5	0.4	0.2	0.5
Th	13.52	13.79	18.77	22.28	24.10	17.82	13.30	19.80	13.73	7.01	10.54	7.99	9.59	0.96	5.73
Hf	5.36	4.91	5.21	8.31	11.85	9.63	7.59	13.70	6.32	5.56	6.87	5.35	10.32	1.06	4.15
Co	1.27	5.81	1.39	0.89	0.81	1.37	1.23	18.80	1.00	3.49	2.79	3.29	23.87	36.85	11.57
Sc	5.16	8.76	2.65	2.85	3.12	1.32	0.98	4.19	3.29	5.23	8.14	11.51	19.41	35.50	16.78
Rb/Sr	1.17	0.99	2.00	7.60	9.62	4.90	4.29	8.92	3.21	0.37	0.78	0.46	0.12	0.06	0.20
ΣREE	215.5	188.6	226.9	374.5	425.9	319.3	283.5	263.5†	307.5	147.8	244.6	199.8	185.6	58.0	187.2
La/Lu	8.90	8.73	5.33	7.50	6.91	6.41	7.83	7.89	6.94	13.74	7.84	8.74	10.44	5.67	8.22
Eu/Sm	0.129	0.269	0.044	0.079	0.051	0.097	0.087	0.102	0.201	0.279	0.284	0.284	0.315	8.00	0.324
Eu/Eu*	0.400	0.388	0.149	0.298	0.167	0.300	0.286	0.332	0.596	0.941	0.722	0.962	1.333	1.50	1.010

*The samples are as follows. See Smith (1978a) for all analyses and sample descriptions.
1. Average of plagioclase bearing rhyolites at the Marquette locality (4 samples).
2. Average of quartz-alkali feldspar bearing rhyolites at the Marquette locality (10 samples).
3. Average of quartz-alkali feldspar bearing rhyolites at the Marcellon locality (5 samples).
4. Montello granite.
5. Granite at Redgranite.
6. Observatory Hill rhyolite.
7. Utley rhyolite.
8. Spring Lake granite of Anderson and others (1980).
9. Flynn's Quarry granite dike.
10. Baxter Hollow granite.
11. Observatory Hill rhyolite dike.
12. Marquette rhyolite dike.
13. Denzer diorite
14. Basalt dike at the Marcellon locality
15. Andesite dike at the Marcellon locality.
†Total does not include Nd

of granophyric granite at Montello and at Redgranite and porphyritic quartz-alkali feldspar rhyolites at Observatory Hill, Berlin, Utley, Taylor Farm, and Endeavor. Finally, group 4 consists of porphyritic quartz-alkali feldspar-plagioclase rhyolites at Marcellon and the rhyolite exposures on the north and south flanks of the Baraboo syncline (Fig. 1). Rocks of the Denzer suite were only briefly discussed by Smith (1978a). Smith (1978a) suggested a cogenetic relationship among the four groups and demonstrated that the granites are the subvolcanic equivalents of the rhyolites. A genetic relationship between the granophyric granites and rhyolites was suggested earlier by Hobbs and Leith (1907) and Asquith (1964), based on the close association of granite and rhyolite outcrops and gross chemical similarities, and by Van Schmus and others (1975) based on similar ages. Anderson and others (1980) also related the rhyolites and granites on the basis of chemical similarity.

Data presented in this paper substantially modify the earlier conclusions of Smith (1978a).

Peraluminous and Metaluminous Suites

Chemical data presented here (Table 2) indicate two major magma types for the south-central Wisconsin rhyolite ash-flow tuffs and granophyric granites. Chemical groups 2 and 4 described by Smith (1978a) are end members of a peraluminous suite, and chemical group 3 represents the differentiated member of a metaluminous suite. A porphyritic granite dike at Flynn's Quarry and the Spring Lake granite of Anderson and others (1980) may represent the other end of the metaluminous suite. The following chemical characteristics help distinguish rocks of the peraluminous suite from those of the metaluminous suite.

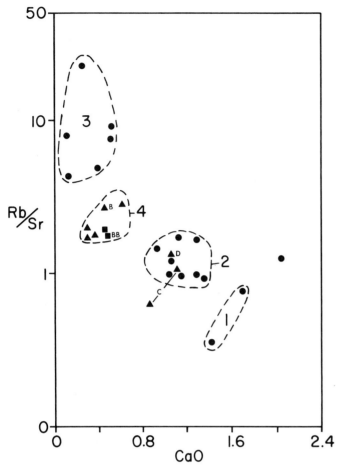

Figure 2. The four chemical groups of Smith (1978a). Letters A through D represent Marcellon rhyolite units. BB is Baraboo rhyolite. Unlabeled data point at about 2% CaO is Marquette unit C sample 91 (See Smith, 1978a).

1. Sr rarely exceeds 60 ppm for metaluminous rocks but ranges from 40 to about 150 ppm for the peraluminous volcanics (Fig. 3a). During differentiation both suites show a decrease in Sr content (Fig. 3g).

2. The maximum Rb/Sr ratio for metaluminous rocks is 23.4, but only 2.7 for peraluminous rhyolites (Fig. 3b).

3. Between SiO_2 values of 71% and 76%, the Ba content shows a stronger negative correlation with SiO_2 in the metaluminous suite than in the peraluminous suite (Fig. 3c).

4. Over the SiO_2 range of 71% to 78%, Rb displays a weak positive correlation with SiO_2 in the metaluminous suite, but remains relatively constant for peraluminous rocks (Fig. 3d).

5. Zr concentration is uniformly higher in the metaluminous suite than in the peraluminous suite (Fig. 3e). In addition, Zr contents are relatively constant within each suite. These data suggest that little or no zircon fractionation occurred. Modeling studies support this observation.

6. Cr concentration varies between 10 and 60 ppm in

the peraluminous suite and shows a positive correlation with SiO_2, but varies between 8 and 38 ppm in the metaluminous suite (Fig. 3f).

7. The peraluminous nature (molecular ratio $Al_2O_3/K_2O + Na_2O + CaO$) of the peraluminous rhyolites increases with differentiation, whereas the opposite is true for the metaluminous rhyolites and granites (Fig. 3h).

8. Both peraluminous and metaluminous suites show light REE enrichment (La/Yb varies between 4.7 and 8.6) and prominent Eu anomalies (Eu/Eu* = 0.15 to 0.4). REE patterns for both suites display relatively flat distributions. However, REE abundances are uniformly higher for the metaluminous rocks (Table 2). Even though there is compositional variation within each suite, the two suites remain compositionally distinct (Fig. 4). Both suites display an increase in all REE (except Eu) with differentiation. This contrasts with the work of Anderson and others (1980) who demonstrated an increase in REE (except Eu) for metaluminous rocks, but a decrease in REE for peraluminous granites.

Baxter Hollow and Denzer Suites

Rocks of the Baxter Hollow suite are peraluminous in character but differ significantly from typical south-central Wisconsin peraluminous rocks in several important respects. The Baxter Hollow granite (the less differentiated member of the suite) is higher in Sr, CaO, Al_2O_3, FeO, Zn, Ba, and Ni and lower in SiO_2, Y, and REE (except La and Ce) than typical peraluminous rhyolites (Figs. 3g, 4). The high Sr, low REE abundances and lack of an Eu anomaly are especially notable. Rocks of the Baxter Hollow suite show significant compositional variation. The Baxter Hollow granite compares with the rhyolite dike of Observatory Hill, the differentiated end member of the Baxter Hollow suite, by having lower Rb/Sr, lower REE abundances, and by lacking an Eu anomaly (Fig. 4).

The Denzer diorite and the tholeiitic basalt and andesite dikes (classified according to Irvine and Baragar, 1971), are characterized by low SiO_2 (48% to 63%) and high Sr (218 to 642 ppm). REE contents vary greatly; most rocks of this suite have enriched light and depleted heavy REE abundances, and they lack an Eu anomaly. Total REE abundances are lower than peraluminous and metaluminous groups but are comparable to rocks of the Baxter Hollow suite (Fig. 5). Notable exceptions to these patterns are basalt dikes at Montello and Marcellon; the Montello dike displays a large negative Eu anomaly and high total REE abundance, whereas the Marcellon dike shows low total REE and lacks an Eu anomaly (Fig. 6). Variations in REE abundances for these dikes may be related to feldspar fractionation and/or contamination by surrounding felsic rocks. The close similarity of the chemistry of this suite to the chemistry of the postulated source material of the pera-

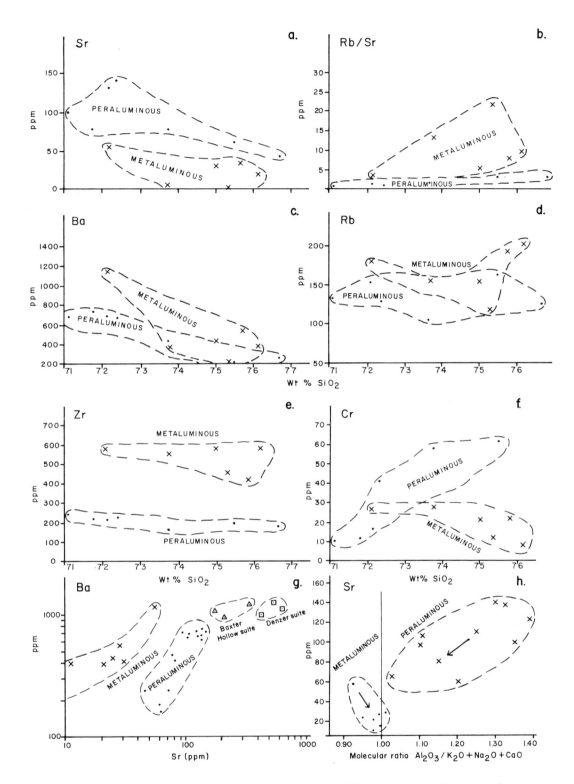

Figure 3. a–f. Harker variation plots showing chemical differences between the peraluminous and metaluminous suites. g. Ba-Sr plot for the four chemical groups. h. Sr-molecular Al_2O_3/K_2O+Na_2O+CaO (the peraluminous index) plot for the peraluminous and metaluminous suites. Peraluminous and metaluminous fields are those defined by Shand (1927).

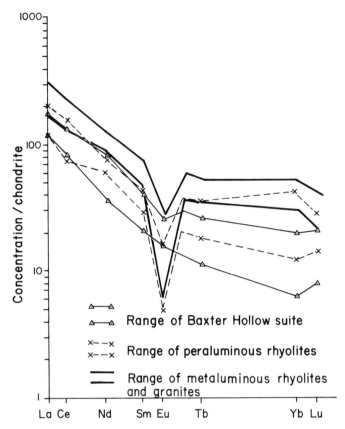

Figure 4. Rare-earth abundances for the peraluminous, metaluminous, and Baxter Hollow suites. The metaluminous rhyolites and granites have a uniformly higher REE content than do the peraluminous rhyolites. The lower curve for the Baxter Hollow suite shows the REE concentration of the Baxter Hollow granite; the upper curve indicates the distribution for the rhyolite dike on Observatory Hill (the differentiated part of the suite).

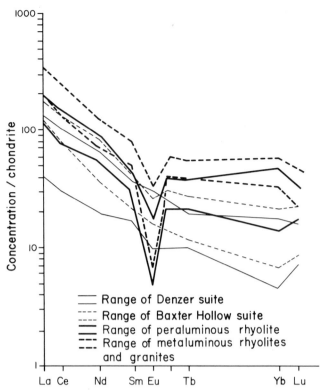

Figure 5. Comparison of the rare-earth abundances for the Denzer suite with the Baxter Hollow suite, the peraluminous suite, and with the metaluminous suite.

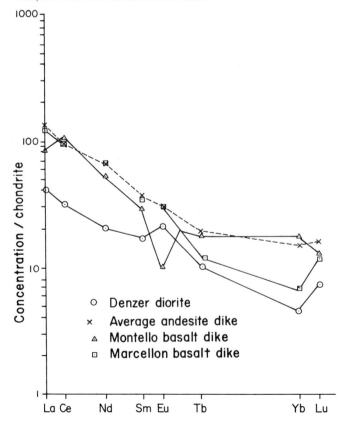

Figure 6. Compositional variation within the Denzer suite.

luminous and metaluminous suites is remarkable. This similarity will be discussed in a later section.

PETROGENETIC MODELS

Geochemical modeling was employed to evaluate the relationships within and between the four rock suites. Geochemical modeling must be used along with other geological data such as field studies and isotopic data for any unique solutions to be determined. Modeling is based on idealized assumptions of initial source composition and proper values of distribution coefficients; therefore, at best, it can only be a crude approximation to reality. Modeling does serve to illustrate the general principles and limiting effects of simple igneous processes (Cox and others, 1979).

Equations used in both partial melting and crystal fractionation are presented in Wood and Fraser (1976), Hanson (1978), Hertogen and Gijbels (1976), and Arth (1976). Major element fractional crystallization is after Wright and Doherty (1970). The behavior of trace elements during partial melting is modeled using the equations of Shaw (1970), who refined techniques developed by Schilling and Winchester (1967) and Gast (1968). Partition coefficients for fractional crystallization and partial melting models are those compiled by Arth (1976). Although partition coefficients may vary with temperature, pressure, and melting proportion (Hertogen and Gijbels, 1976), constant coefficients are used for all modeling calculations in the present study.

Peraluminous and Metaluminous Suites

The common spatial association and age of the peraluminous and metaluminous suites suggest that they are cogenetic and possibly comagmatic and that they are related by a simple process such as partial melting or fractional crystallization. Other models involving contamination and residual phases can be immediately ruled out on the basis of rock petrography and chemistry. There are many chemical similarities between the two suites. For example, both show enriched light REE patterns and a moderate depletion in the heavy REE. Even though there is compositional variation within each suite, there are small but constant differences between the suites (Figs. 3, 4). If one suite were derived from the other by fractional crystallization, the more differentiated suite should be enriched in elements such as Si, K, Rb, and REE, except Eu, and depleted in Ti, Al, Fe, Mg, Ca, Cr, and Zr. If we assume that the undifferentiated part of the metaluminous suite was derived by fractional crystallization of the undifferentiated part of the peraluminous suite, then the process must have occurred without appreciably enriching the differentiate in Si or Rb or depleting it in Cr or Zr. Also Eu contents of the metaluminous suite (the differentiate) are higher than that of the peraluminous suite (the parent). These chemical characteristics preclude a simple relationship by fractional crystallization *between* the two suites. This conclusion was supported by modeling studies. Many models were developed that related the suites to a common parent or derived one from the other, but the models either showed moderate to high residuals between actual and model differentiates or produced fractional mineral proportions that were not consistent with actual liquidus phase concentrations. Alternatively, the two suites may have formed individually by partial melting of crustal material. This hypothesis was tested by modeling studies.

The undifferentiated part of the peraluminous suite is similar in major and minor element chemistry to the undifferentiated peraluminous post-Penokean granites described by Anderson and others (1980). They have shown that 10% fusion of tonalitic to granodioritic crustal material could have formed the undifferentiated peraluminous granites. In addition, Anderson and Cullers (1978) suggested that 20% fusion of tonalitic to granodioritic lower crust may have

produced the undifferentiated parts of the 1.5-b.y.-old Wolf River Batholith in northeastern Wisconsin. A similar tonalitic to granodioritic source seems likely for the peraluminous suite of rhyolites reported here. Model studies show that a tonalitic crustal source having the composition 60% plagioclase, 25% quartz, 5% orthoclase, and 10% biotite is the most likely parent or the peraluminous suite. The range of published compositions for tonalite, granodiorite, and andesite in Anderson and Cullers (1978) was used in these models (Fig. 7a). The Marquette rhyolite, the undifferentiated member of the suite, can be generated by 16% nonmodal fractional melting of plagioclase, quartz, and orthoclase in the proportions 48/28/24 from a crustal source having the above-mentioned composition (Fig. 7a). The nonmodal melting proportions are consistent with the average normalized mode of the Marquette rhyolite (plagioclase 46%, quartz 22%, orthoclase 32%).

Similar modeling was done for the metaluminous suite, assuming that the undifferentiated member of this suite is represented by the average of the composition of Flynn's Quarry granite dike and the Spring Lake granite of Anderson and others (1980). Sixteen percent nonmodal fractional melting of plagioclase, quartz, orthoclase, biotite, apatite, and zircon in the proportions 8/27/55/9/0.1/0.1 from a tonalitic parent containing 55% plagioclase, 30% quartz, 10% orthoclase, 5% biotite, 0.3% apatite, and 0.1% zircon will yield melts with approximately the same composition as the metaluminous granites and rhyolites (Fig. 7b). Nonmodal melting proportions are consistant with the mode of the Spring Lake granite and with the normalized mode of the Flynn's Quarry dike (Table 1). This model implies that the source material for the metaluminous suite is higher in REE than the average source material for the peraluminous suite.

In conclusion, these modeling studies suggest that both the metaluminous and peraluminous suites were derived from separate intermediate composition sources. These sources vary slightly but significantly in trace element content and mineralogy. Our model studies also indicate that melting of the same parent to different degrees will not produce melts having satisfactory matches to the composition of the two suites. This conclusion can also be reached qualitatively. Since the less differentiated rocks of each suite are different, it follows that the source material melted to produce each should also be dissimilar. This conclusion assumes that the composition of the less differentiated member of each suite approximates that of the parent magma.

The compositional variation *within* the two suites may be explained by fractional crystallization. To test this hypothesis the major elements REE, Rb, Sr, and Ba were used to model fractional crystallization. Mineral compositions used in the models are listed in Table 3. Phases separated are realistic in terms of rock mode. Modeling of fractional

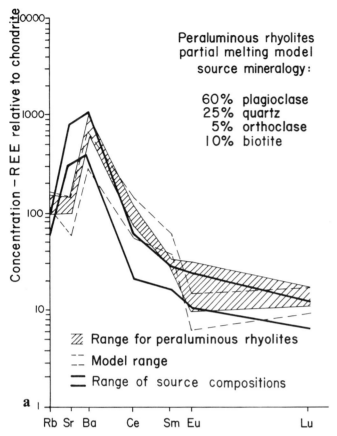

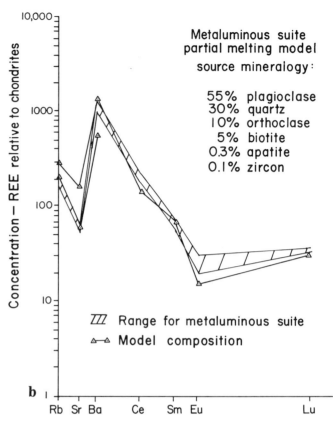

Figure 7. a. Model for the formation of the peraluminous suite by the nonmodal fractional melting of plagioclase, quartz, and orthoclase in the proportions 48/28/24 from a tonalitic parent. Range of published compositions for tonalite, granodiorite, and andesite (Anderson and Cullers, 1978) used in the partial melting models is shown in solid lines. b. Model for the formation of the metaluminous suite by the nonmodal fractional melting of plagioclase, quartz, orthoclase, biotite, apatite, and zircon in the proportions 8/27/55/9/0.1/0.1 from a quartz diorite source. For Rb, Sr, and Ba the range of source compositions on Figure 9 were used for modeling, but for REE only the higher limit of source concentrations (fig. 9) was used.

crystallization was done in two steps. First a model was derived by using major elements, and then it was refined by using the trace elements and REE.

The variation of major elements within the peraluminous suite suggests that feldspar and a mafic phase were removed from the differentiated member of this suite; however, the drop in the peraluminous index (molecular ratio $Al_2O_3/K_2O + Na_2O + CaO$) from 1.40 to 1.02 (Fig. 3h) suggests that an aluminous-rich phase such as biotite or muscovite was extracted. In addition, the negative correlation of Ba with SiO_2 suggests that a phase that concentrates Ba, such as biotite, was withdrawn. Our modeling studies confirm this observation by suggesting that substantial (5%) fractionation of biotite occurred. The relatively constant Zr content (Fig. 3e) indicates that large amounts of zircon were not extracted. The differentiation of the peraluminous rhyolites using major element variation is approximated by 19.9% fractionation of plagioclase (10.9%), alkali feldspar (7.1%), and biotite (1.9%) (Table 4). A refined model using REE, Rb, Sr, and Ba requires 23.1%

fractionation of plagioclase (11%), alkali feldspar (7.1%), and biotite (5%) (Fig. 8a).

The peraluminous index for the metaluminous suite increases from 0.92 to 1.01 (Fig. 3h), suggesting that a metaluminous phase such as hornblende was removed. Major element modeling confirms this observation. The variation within the metaluminous suite can be approximated by 16.6% fractionation of plagioclase (0.6%), alkali feldspar (11.3%), and hornblende (4.7%) (Table 5). A refined model using REE, Rb, Sr, and Ba requires 26% fractionation of plagioclase (5%), alkali feldspar (15%), biotite (3%), and hornblende (3%) (Fig. 8b). It should be emphasized that neither model produces a unique solution.

Baxter Hollow Granite

The Baxter Hollow granite was thought by Smith (1978a) to represent the source material for the other south-central Wisconsin granites and rhyolites. Present modeling studies show, however, that the peraluminous and meta-

TABLE 3. MINERAL COMPOSITIONS USED IN DIFFERENTIATION MODELS

Oxides	Plagioclase*	Alkali Feldspar[+]	Biotite[ξ]	Hornblende
SiO_2	60.00	67.27	34.75	40.76
TiO_2	-	-	3.42	1.62
Al_2O_3	26.00	18.35	14.10	9.31
Fe_2O_3	-	-	-	-
FeO	-	0.92	31.66	29.61
MgO	-	-	2.74	2.72
CaO	6.00	0.15	0.49	10.44
Na_2O	8.00	6.45	0.02	1.76
K_2O	-	7.05	0.95	1.44

*Estimated for An_{36}.

[+]Alkali feldspar from Deer and others (1966).

[ξ]Biotite and hornblende from Anderson and Cullers (1978).

TABLE 4. COMPARISON OF ACTUAL AND PREDICTED DATA FOR DIFFERENTIATION WITHIN THE PERALUMINOUS SUITE

Oxides	Parent*	Actual Differentiate[+]	Model Differentiate	Error
SiO_2	72.84	75.78	75.72	-0.20
TiO_2	0.30	0.23	0.27	0.03
Al_2O_3	12.87	12.14	12.25	0.06
Fe_2O_3 FeO	3.19	2.06	2.03	-0.03
MgO	0.09	0.12	0.05	-0.14
CaO	1.09	0.40	0.62	0.18
Na_2O	3.15	3.72	2.78	-0.79
K_2O	6.18	5.44	6.39	0.78

Percent Fractionated Minerals	
PLAG	11.9
KSPR	7.1
HB	1.9
Total Fractionation	19.9
% Melt	80.1

*Average of the Marquette Rhyolite.

[+]Average of the Marcellon Rhyolite.

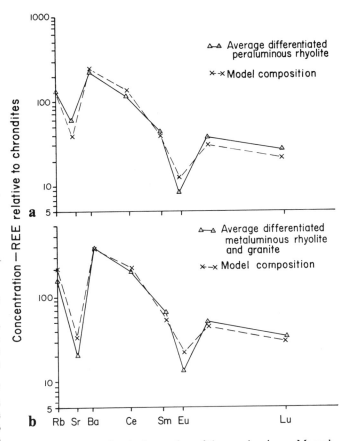

Figure 8. a. Model for the formation of the peraluminous Marcellon rhyolites by the fractional crystallization of plagioclase (11%), alkali feldspar (7.1%), and biotite (5%) from the Marquette rhyolite. b. Model for the formation of the metaluminous rhyolites and granophyric granites by the fractional crystallization of plagioclase (5%), alkali feldspar (15%), biotite (3%), and hornblende (3%) from a magma having the average composition of the granite dike at Flynn's Quarry and the Spring Lake granite of Anderson and others (1980).

luminous magma types cannot be generated from the Baxter Hollow granite by either partial melting or fractional crystallization. Models either require mineral proportions that are not consistant with modal mineralogy or that show large residuals between model and actual composition. Also, the Baxter Hollow suite is not depleted in REE when compared to the peraluminous suite (Fig. 5), as would be expected if the two rock types were related by fractional crystallization or partial melting. The Baxter Hollow suite consists of the Baxter Hollow granite as the undifferentiated end member of the suite and the Observatory Hill and Marquette rhyolite dikes as differentiated end products. The suite represents an independently derived magma series. The following evolutionary models for the formation of the Baxter Hollow granite were tested: (1) formation by fractional crystallization of an interme-

diate composition magma and (2) evolution by partial melting of either crustal or mantle material.

Fractional Crystallization. The Baxter Hollow granite may be derived from a diorite magma (60% plagioclase, 40% hornblende) by fractionation of 10% plagioclase. Trace element data support this model (Fig. 9), but the best fit model using major elements (the fractionation of 10% plagioclase, 5% hornblende, and 1% magnetite) results in large differences between model and actual compositions. Also, there are severe volumetric problems to this model, since no large coeval diorite intrusion crops out or is observed in the subsurface. The Denzer diorite located 20 km to the west (Fig. 1) is the only rock of the appropriate composition, but apparently it is a small pluton (as indicated by the study of well cuttings; Smith, 1978b) and as a result is unsuitable as a source.

Partial Melting. The high Sr and lack of an Eu anomaly for the Baxter Hollow granite require either that plagio-

TABLE 5.   COMPARISON OF ACTUAL AND PREDICTED DATA FOR DIFFERENTIATION
WITHIN THE METALUMINOUS SUITE

Oxides	Parent*	Actual Differentiate[†]	Model Differentiate	Error
SiO₂	72.76	75.99	75.96	0.01
TiO₂	0.28	0.14	0.27	0.10
Al₂O₃	14.28	12.43	12.42	0.00
Fe₂O₃ FeO	2.39	2.11	2.10	-0.01
MgO	0.37	0.35	0.39	0.03
CaO	1.23	0.52	0.54	0.01
Na₂O	3.96	3.39	3.41	0.01
K₂O	4.48	4.91	4.92	0.01

Precent Fractionated Minerals	
PLAG	0.6
KSPR	11.3
BIOT	4.7
Total Fractionation	16.6
% Melt	84.4

*Average of the Flynn's Quarry Granite and the Spring Lake Granite of
Anderson and others (1980).

†Average of the metaluminous granophyric granites and quartz-alkali
feldspar rhyolites.

clase was not an equilibrium phase during partial melting or that a large degree of melting of an intermediate composition crustal source occurred; moreover, the low REE abundances suggest that mineral phases that retain REE in the residuum, such as hornblende or garnet, were in equilibrium with the melt during partial melting.

The Baxter Hollow granite may be derived from a garnet-bearing mantle source such as eclogite or garnet peridotite by 1% modal fractional melting (Fig. 10). The model source is assumed to be enriched in REE 10 times over that of chondrites. Rb, Sr, and Ba concentrations for eclogite and peridotite are from the compilations of Wedepohl and Muramatsu (1979) and Wedepohl (1978). The composition of the Baxter Hollow granite produced by 1% melting of a mantle source agrees well with the actual trace element abundances, except that the model rock displays a positive Eu anomaly and low Ce and Sm concentrations (Fig. 10). But, according to Wyllie (1976), it is unlikely that granites or rhyolites are primary magmas derived from mantle peridotite.

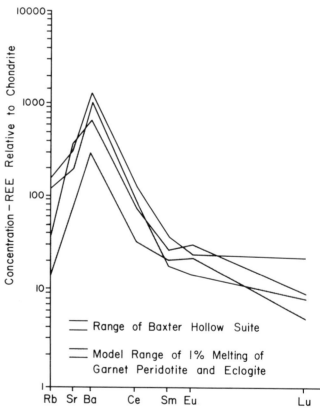

Figure 10. Models for the formation of the Baxter Hollow granite by the modal fractional melting of eclogite and garnet peridotite source rocks. Source is assumed to be enriched in REE 10 times over chondrite abundances. Eclogite mineralogy used for modeling varies from 95% clinopyroxene, 5% garnet to 98% clinopyroxene, 2% garnet. Garnet peridotite varies in mineralogy from 52% olivine, 17% orthopyroxene, 12% clinopyroxene, 14% garnet to 10% olivine, 30% orthopyroxene, 50% clinopyroxene, and 5% garnet. All garnet peridotite model rocks contain 5% phlogopite.

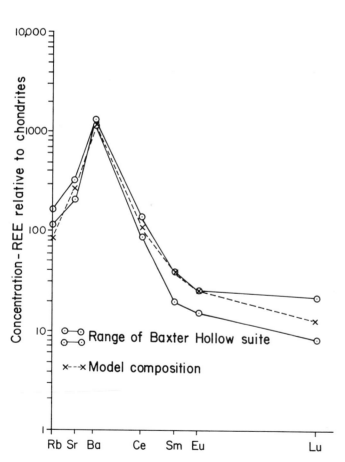

Figure 9. Model for the formation of the Baxter Hollow granite by the fractional crystallization of 10% plagioclase from a diorite source.

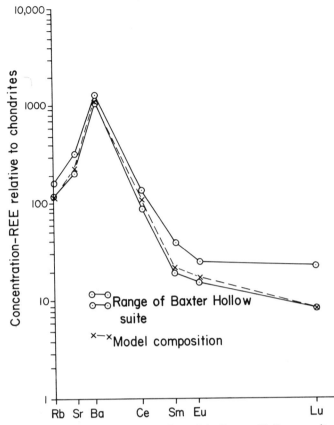

Figure 11. Model for the formation of the Baxter Hollow granite by the modal batch melting of a diorite source rock.

The Baxter Hollow granite also may be produced by 66% modal batch melting or a diorite (60% plagioclase, 40% hornblende) source (Fig. 11). The large degree of melting proposed by this model is bothersome, since large-scale melting of source rock does not seem to be a common occurrence during the 1.76-b.y.-old event. Model studies described here suggest 16% partial melting for coeval rocks. Moreover, Anderson and others (1980) proposed 10% melting of an intermediate composition parent for the 1.76-b.y.-old peraluminous and metaluminous granites, and Anderson and Cullers (1978) suggest 20% melting to produce parts of the 1.5-b.y.-old Wolf River batholith. A locally intense thermal event would be necessary to derive the Baxter Hollow granite from a dioritic source by partial melting. Perhaps the heat necessary to cause such large-scale melting was provided by Denzer suite basaltic magma rising through the crust from a mantle source (see next section).

Although a unique solution cannot be obtained, the available data marginally support the origin of the Baxter Hollow granite by the partial melting of an intermediate composition crustal parent. Isotopic data, especially initial $^{87}Sr/^{86}Sr$, are required to choose between models. Even though there are uncertainties in source composition, it is clear that the Baxter Hollow granite represents a part of a rock suite formed independently of the peraluminous and metaluminous suites during the 1.76-b.y.-old event.

Variation within the Suite. The rhyolite dikes on Observatory Hill and at the Marquette locality may have formed from a magma with the composition of the Baxter Hollow granite by fractional crystallization. By using the major elements, a model involving the fractionation of 1% plagioclase, 15% alkali feldspar, 20% quartz, and 1% biotite (37% total fractionation) from the Baxter Hollow magma is required to produce the rhyolite dikes (Table 6). A refined model using REE, Rb, Sr, and Ba confirms these percentages (Fig. 12).

Denzer Suite

To determine the origin of the Denzer suite, two evolutionary scenarios were modeled: (1) origin of the tholeiitic basalt dikes (the less differentiated part of the suite) as a residual phase from the melting of the peraluminous or metaluminous suites and (2) derivation by partial melting of a mantle source.

Partial melting of a quartz diorite parent to produce the metaluminous and peraluminous suites will leave a residuum of approximately the same composition as the Denzer suite rocks. However, the model residuum shows a prominent positive Eu anomaly. This is not a characteristic of the Denzer suite in general. Also, rocks of the suite do not display field or petrographic characteristics of residual rocks.

To determine the most appropriate mantle source material for the Denzer suite several possible parents must be evaluated. According to Wyllie (1979), the upper mantle is heterogenous and is composed of peridotite with olivine and two pyroxenes with an aluminous mineral such as plagioclase, spinel, or garnet (depending on the depth within

TABLE 6. COMPARISON OF ACTUAL AND PREDICTED DATA FOR DIFFERENTIATION WITHIN THE BAXTER HOLLOW SUITE

Oxides	Parent	Actual Differentiate	Model Differentiate	Error
SiO_2	71.14	70.54	70.54	0.00
TiO_2	0.30	0.44	0.56	-0.13
Al_2O_3	14.23	14.48	14.51	-0.03
Fe_2O_3	3.56	4.00	3.98	0.02
FeO				
MgO	0.91	0.49	0.37	0.12
CaO	1.70	1.09	1.18	-0.09
Na_2O	4.31	4.71	4.53	0.18
K_2O	3.71	3.47	3.54	-0.08

Percent Fractionated Minerals	
PLAG	1.0
KSPR	15.0
QUARTZ	20.0
BIOTITE	1.0
Total Fractionation	37.0
% Melt	63.0

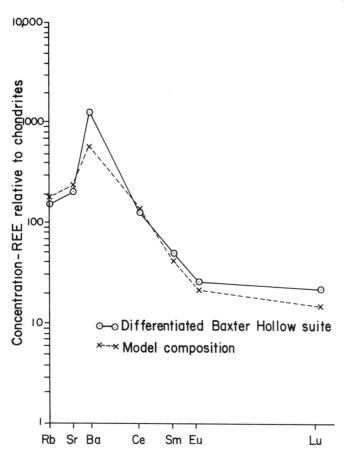

Figure 12. Model for the formation of the Observatory Hill and Marquette rhyolite dikes by fractional crystallization of plagioclase (1%), alkali feldspar (15%), quartz (20%), and biotite (1%) from the Baxter Hollow granite.

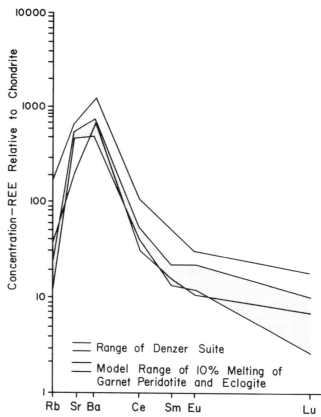

Figure 13. Model for the formation of the Denzer diorite by the modal fractional melting of eclogite and garnet peridotite source rocks. Source material is assumed to be enriched in REE 10 times over that of chondrites. Mineralogical variation of mantle source rocks is the same as that for the Baxter Hollow granite models (see Fig. 10).

the mantle). Partial melting of a spinel-bearing peridotite containing 10% clinopyroxene, 30.1% orthopyroxene, 56.4% olivine, and 3.5% spinel and having chondritic REE abundances produces upon 5% partial melting a liquid with approximately 15 times chondrite abundance for each REE (Mysen, 1979). Arth and Hanson (1975) modeled the melting of a plagioclase-bearing peridotite enriched in REE 3 times over that of chondrites with the composition 63% olivine, 15% orthopyroxene, 7% clinopyroxene, 7% anorthite, and 3% albite. Upon 10% melting, a liquid containing approximately 30 times chondrite abundance for each REE was derived. In summary, melting of mantle rock without garnet produces relatively flat REE patterns that are quite unlike the distribution for the Denzer suite tholeiitic basalts. The higher total REE and the higher La/Lu ratio of Denzer suite basalts require that garnet is present in the residuum (see model studies reported in Yoder, 1976).

The present modeling studies suggest that the undifferentiated parts of the Denzer suite may have been derived by 10% partial melting of either garnet peridotite or eclogite at depth within the upper mantle (Fig. 13). The mantle source is assumed to be enriched in REE 10 times over that of

chondrites. The models for garnet peridotite varied in composition from 15% olivine, 30% orthopyroxene, 50% clinopyroxene, and 5% garnet to 52% olivine, 17% orthopyroxene, 12% clinopyroxene, and 14% garnet, and for eclogite between 95% clinopyroxene, 5% garnet, and 98% clinopyroxene, 2% garnet. Eclogite and pyroxene-rich garnet peridotite provided the best fit between model and actual rocks. The relatively high Rb and Ba contents of the Denzer suite rocks implies that the source rock contains a phase such as phlogopite or amphibole that would concentrate these elements in the liquid upon small amounts of fusion. As little as 5% phlogopite and/or amphibole in the model source provides a good fit to the actual rock composition (Fig. 13).

If the tholeiitic basalts are assumed to be the parent magma of the Denzer suite, then the Denzer diorite and the andesite dikes may have been derived by fractional crystallization of this parent. However, modeling studies were not successful in substantiating this hypothesis. Compositional zoning within individual intrusions and contamination may be complicating factors that preclude the success of simple models.

The Denzer suite has compositional similarities to the model Penokean-aged source rocks for the peraluminous and metaluminous granites and rhyolites (Fig. 14). This similarity may not be a coincidence, but may indicate that the Penokean-aged intermediate rocks and the post-Penokean diorites and mafic dikes formed from similar source material in the upper mantle. However, the volume of intermediate composition magma generated during the post-Penokean event was considerably less than was produced 60 m.y. earlier, during the Penokean orogeny.

ANOROGENIC NATURE OF THE ROCK SUITES AND SUMMARY

The Precambrian rhyolites and granites of south-central Wisconsin are unique in that they are the only exposures in the Lake Superior region of a 1.76-b.y.-old volcanic-plutonic province. These rocks were formed in the interval between the Penokean orogeny, which involved extensive igneous activity in northern and central Wisconsin, and the anorogenic emplacement of the Wolf River batholith. The peraluminous and metaluminous suites of rhyolites and granites probably formed by partial melting of similar intermediate composition crustal parents. Fractional crystallization produced the limited compositional variation observed within each suite. The Denzer suite was probably derived by a small amount of melting of a mantle source composed of garnet peridotite or eclogite. The Baxter Hollow suite was perhaps produced by large-scale melting of a crustal source. Heat generated by the rise of Denzer suite magma may have caused the local large-scale melting necessary to produce the Baxter Hollow magma from a dioritic source. Baxter Hollow and Denzer suite magmas were emplaced into the peraluminous and metaluminous granites and rhyolites as small stocks and dikes.

The tectonic framework of the 1.76-b.y.-old rocks can be determined by careful study of the chemical data. In general, rock suites emplaced during an anorogenic event are higher in SiO_2 and P_2O_5, higher in alkalies (especially K), low in CaO, Al_2O_3 and MgO, and show definite iron enrichment when compared to orogenic rock associations (Martin and Piwinskii, 1972). Also, in anorogenic suites few rocks have SiO_2 contents that are less than 65% (Rogers and Greenberg, 1981). When plotting $\Sigma FeO/\Sigma FeO +$ MgO for anorogenic and orogenic suites, ratios of 0.8 to 1.0 are characteristic of anorogenic suites and ratios of 0.5 to 0.8 for syntectonic suites (Anderson and Cullers, 1978). The mean $\Sigma FeO/\Sigma FeO +$ MgO for granites and rhyolites of south-central Wisconsin is 0.9; the rocks are clearly anorogenic. These rocks are also too high in alkalies and too low in CaO, Al_2O_3, and MgO to be synorogenic. Moreover, they plot in the alkali granite field on a $\log_{10} (K_2O/MgO)$ versus SiO_2 plot (Ghuma and Rogers, 1978) and are typical of the rhyolites and granites that form shortly following a

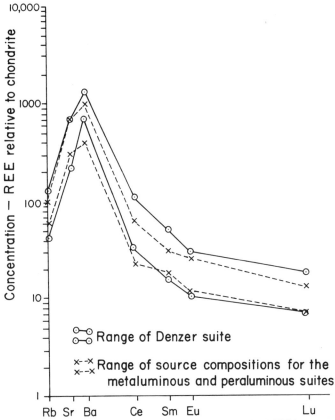

Figure 14. Close similarity in Rb, Sr, Ba, and the REE between the model source for the metaluminous and peraluminous rocks and the Denzer suite.

major orogenic event (Rogers and Greenberg, 1981). Anderson and others (1980) arrived at similar conclusions by their study of the 1.76-b.y.-old granites in south-central and northern Wisconsin.

ACKNOWLEDGMENTS

I especially thank John D. Jones and Ward Rigot of the Phoenix Memorial Laboratory, University of Michigan, for completing the REE analyses. Financial support by the Department of Energy (D.O.E.) to the Phoenix Memorial Laboratory through the University Research Reactor Assistant Program and Reactor Facility Cost Sharing Program made this work possible. Thanks are due to D. L. Weide and his cartography students for carefully drafting the figures, and to Ruth Roark for typing the final and earlier versions of this paper. Diane Pyper Smith made many helpful editorial corrections.

REFERENCES CITED

Anderson, J. L., and Cullers, R. L., 1978, Geochemistry and evolution of the Wolf River Batholith, a late Precambrian rapakivi massif in northern Wisconsin, U.S.A.: Precambrian Research, v. 7, p. 287–324.

Anderson, J. L., Cullers, R. L., and Van Schmus, W. R., 1980, Anorogenic metaluminous and peraluminous granite plutonism in the mid-Proterozoic of Wisconsin, USA: Contributions to Mineralogy and Petrology, v. 74, p. 311–328.

Arth, J. G., 1976, Behavior of trace elements during magmatic processes—a summary of theoretical models and their applications: U.S. Geological Survey Journal of Research, v. 4, p. 41–47.

Arth, J. G., and Hanson, G. N., 1975, Geochemistry and origin of the early Precambrian crust of northern Minnesota: Geochimica et Cosmochimica Acta, v. 39, p. 324–362.

Asquith, G. B., 1964, Origin of Precambrian Wisconsin rhyolites: Journal of Geology, v. 72, p. 835–847.

Banks, P. O., and Rebello, D. P., 1969, Zircon ages of a Precambrian rhyolite, northeastern Wisconsin: Geological Society of America Bulletin, v. 80, p. 907–910.

Brandon, C. N., Smith, E. I., and Luther, F. R., 1980, the Precambrian Waterloo Quartzite, southeastern Wisconsin: Evolution and significance [abs.]: Proceedings and Abstracts 26th Annual Institute on Lake Superior Geology, p. 17–18.

Cox, K. G., Bell, J. D., and Pankhurst, R. J., 1979, The interpretation of igneous rocks: London, George Allen and Unwin, 450 p.

Cudzilo, T. F., 1978, Geochemistry of early Proterozoic igneous rocks, northeastern Wisconsin and Upper Michigan [Ph.D. thesis]: Lawrence, University of Kansas, 194 p.

Dalziel, I.W.D., and Dott, R. H., 1970, Geology of the Baraboo district, Wisconsin: Wisconsin Geological and Natural History Survey Information Circular 14, 164 p.

Deer, W. A., Howie, R. A., and Zussman, J., 1966, An introduction to the rock forming minerals: London, Longman Group Limited, 528 p.

Gast, P. W., 1968, Trace element fractionation and origin of tholeiitic and alkaline magma types: Geochimica et Cosmochimica Acta, v. 32, p. 1057–1086.

Gates, R. M., 1942, The Baxter Hollow granite cupola: American Mineralogist, v. 27, p. 699–711.

Geiger, C. A., Petro, W. L., and Guidotti, C. V., 1982, Some aspects of the petrologic and tectonic history of the Precambrian rocks of Waterloo, Wisconsin: Geoscience Wisconsin, v. 6, p. 21–40.

Ghuma, M. A., and Rogers, J.J.W., 1978, Geology, geochemistry, and tectonic setting of the Ben Ghnema batholith, southern Libya: Geological Society of America Bulletin, v. 89, p. 1351–1358.

Goldich, S. S., Nier, A. D., Baadsgaard, H., Hoffman, J. H., and Krueger, H. W., 1961, The Precambrian geology and geochronology of Minnesota: Minnesota Geological Survey Bulletin 41, 193 p.

Green, J. C., 1977, Keweenawan plateau volcanism in the Lake Superior region *in* Baragar, W.R.A., and others, eds., Volcanic regimes in Canada: Geological Association of Canada Special Paper 16, p. 407–422.

Hanson, G. H., 1978, The application of trace elements to the petrogenesis of igneous rocks of granitic composition: Earth and Planetary Science Letters, v. 38, p. 26–43.

Hertogen, J., and Gijbels, R., 1976, Calculations of trace element fractionation during partial melting: Geochimica et Cosmochimica Acta, v. 40, p. 313–322.

Hobbs, W. H., and Leith, C. K., 1907, The Precambrian volcanic and intrusive rocks of the Fox River Valley, Wisconsin: University of Wisconsin Bulletin 158, Science Series, v. 3, p. 247–277.

Irvine, T. N., and Baragar, W.R.A., 1971, A guide to the chemical classification of the common volcanic rocks: Canadian Journal of Earth Science, v. 8, p. 523–548.

Martin, R. F., and Piwinskii, A. J., 1972, Magmatism and tectonic setting: Journal of Geophysical Research, v. 70, p. 3485–3496.

Mysen, B. O., 1979, Trace-element partitioning between garnet peridotite minerals and water-rich vapor: Experimental data from 5 to 30 kbar: American Mineralogist, v. 64, p. 274–287.

Petro, W. L., 1980, Mineralogy and chemistry of middle Precambrian

(Xg) granite plutonic rocks from northern Wisconsin: Proceedings and Abstracts 26th Annual Institute on Lake Superior Geology, p. 70.

Rogers, J.J.W., and Greenberg, J. K., 1981, Trace elements in continental-margin magmatism: Part III. Alkali granites and their relationship to cratonization: Geological Society of America Bulletin; Part I, v. 92, p. 6–9; Part II, v. 92, p. 57–93.

Schilling, J. G., and Winchester, J. W., 1967, Rare earth fractionation and magmatic processes *in* Runcorn, S. K., ed., Mantles of the Earth and terrestrial planets: New York, Interscience, p. 267–283.

Shand, S. J., 1927, Eruptive rocks: London, Thomas Murby and Co., 360 p.

Shaw, D. M., 1970, Trace element fractionation during anatexis: Geochimica et Cosmochimica Acta, v. 34, p. 237–243.

Sims, P. K., 1976, Precambrian tectonics and mineral deposits, Lake Superior region: Economic Geology, v. 71, p. 1092–1127.

Smith, E. I., 1978a, Precambrian rhyolites and granites in south-central Wisconsin: Field relations and geochemistry: Geological Society of America Bulletin, v. 89, p. 875–890.

Smith, E. I., 1978b, Introduction to Precambrian rocks of south-central Wisconsin: Geoscience Wisconsin, v. 2, p. 1–17.

Smith, E. I., and Stupak, W. A., 1978, A Fortran IV program for the classification of volcanic rocks using the Irvine and Baragar classification scheme: Computers and Geoscience, v. 4, p. 89–99.

Stark, J. T., 1932, Igneous rocks of the Baraboo district, Wisconsin: Journal of Geology, v. 40, p. 119–139.

Van Schmus, W. R., 1976, Early and middle Proterozoic history of the Great Lakes area, North America: Philosophical Transactions of the Royal Society of London, ser. A, v. 280, p. 605–628.

Van Schmus, W. R., 1978, Geochronology of the southern Wisconsin rhyolites and granites: Geoscience Wisconsin, v. 2, p. 19–24.

Van Schmus, W. R., 1980, Chronology of igneous rocks associated with the Penokean orogeny in Wisconsin: Geological Society of America Special Paper 182, (Goldich Volume), p. 159–168.

Van Schmus, W. R., Thurman, M. E., and Peterman, Z. E., 1975, Geology and Rb-Sr chronology of middle Precambrian rocks in eastern and central Wisconsin: Geological Society of America Bulletin, v. 86, p. 1255–1265.

Van Schmus, W. R., and Bickford, M. E., 1981, Proterozoic chronology and evolution of the midcontinent region, North America, *in* Kröner, A., ed., Precambrian plate tectonics: Elsevier, p. 261–296.

Wedepohl, K. H., 1978, Section 73-E. Abundance in common igneous rock types, crustal abundance: Handbook of Geochemistry, v. 2, Berlin, Springer.

Wedepohl, K. H., and Muramatsu, Y., 1979, The chemical composition of kimberlites compared with the average composition of the three basaltic magma types, *in* Boyd, F. R., and Meyer, H.O.A., Kimberlites, diatremes and diamonds, their geology, petrology and geochemistry: Proceedings of the Second International Kimberlite Conference, American Geophysical Union, p. 300–312.

Wood, B. J., and Fraser, D. G., 1976, Elementary thermodynamics for geologists: London, Oxford University Press.

Wright, T. L., and Doherty, P. D., 1970, A linear programming and least squares computer method for solving petrologic mixing problems: Geological Society of America Bulletin, v. 81, p. 1995–2008.

Wyllie, P. J., 1979, Magmas and volatile components: American Mineralogist, v. 64, p. 464–500.

Yoder, H. S., 1976, Generation of basaltic magma: Washington, National Academy of Sciences, 265 p.

MANUSCRIPT ACCEPTED BY THE SOCIETY MARCH 4, 1983

Geological Society of America
Memoir 160
1983

The Proterozoic red quartzite enigma in the north-central United States: Resolved by plate collision?

R. H. Dott, Jr.
Department of Geology and Geophysics
University of Wisconsin
Madison, Wisconsin 53706

ABSTRACT

Prepaleozoic red quartzites in the north-central United States are among the world's oldest red beds. Not only do they constitute evidence about Proterozoic weathering and the advent of significant atmospheric free oxygen but they also provide important evidence bearing upon the sedimentary and tectonic history of North America. All of the red quartzites postdate the Penokean orogeny (1,760 to 1,860 m.y., broadly defined), and isotopic evidence permits that some could conceivably be as young as 1,200 m.y. (Keweenawan basalt eruption). Although these different quartzites may not be strictly correlative, it is probable that those south of Lake Superior (Baraboo, Sioux, Barron) fall within the narrower span of 1,450 to 1,750 m.y., here called the *Baraboo interval* for the best-dated sequence of sedimentation, deformation, and metamorphism. It has been suggested that their deposition probably predates a major 1,615 to 1,630-m.y. Rb-Sr resetting event, which would narrow the depositional age span considerably more. The severity of deformation of the quartzites is enigmatic in terms of their maturity and present intracratonic location.

Sources of the red quartzitic sediments were quartz-bearing silicic volcanic rocks, such as underlie the Baraboo Quartzite, and older sedimentary, plutonic, and minor metamorphic rocks. Predominance of aluminous kaolinite and pyrophyllite in fine red argillaceous strata and near-absence of feldspar in the sandstones suggest mature chemical weathering in the source. Without vegetation to stabilize soils, this seems possible only on a stable landscape with little topographic relief and in a warm, humid climate. Braided rivers apparently drained the hinterland and formed a dominantly sandy coastal plain. Marine transgression may have begun during deposition of the sandstones, for the youngest strata preserved, which are black slates and iron-formation–bearing dolomite overlying the Baraboo Quartzite, are considered to be marine.

Deposition of a red, mature quartzose (90% to 95%) sedimentary wedge up to 2,000 m thick points to a stable, passive continental margin on the southern edge of a Proto–North American craton. Later deformation was intense and greenschist metamorphism was pervasive, implying compressive orogenesis. The hypothesis is proposed that a major, hitherto unrecognized plate suture exists beneath Iowa and Illinois. Southward subduction during or after deposition of the red quartzites caused consumption of a sea floor that lay to the south of Proto–North America and culminated in collision either of a volcanic island arc or of another continent with southern Proto–North America about 1,615 to 1,630 m.y. ago. Foliated granitic plutons 1,625 m.y. old are known in northern Kansas and Missouri, and the Mazatzal belt of Arizona and New Mexico that contains plutonic and volcanic rocks of

1,610 to 1,680-m.y. ages also seems to be related to some such event. Apparently emplacement of the anorogenic Wolf River batholith in Wisconsin (about 1,500 m.y.) and eruption of widespread rhyolites to the south (1,380 to 1,480 m.y.) postdated the collision. Whatever may be the merits of this particular hypothesis, the red quartzites provide important constraints upon concepts of North American continent-building during Proterozoic time.

INTRODUCTION

Prepaleozoic red quartzites have long been known in the north-central United States (Fig. 1), where they have been quarried extensively—first by northern plains Indians for making ceremonial pipes from red argillite (pipestone) and later by their successors for paving stone, railroad ballast, and abrasives. Those that fall in the 1,450 to 1,750-m.y. age span seem to comprise a hitherto little-recognized and unnamed stratigraphic division, which is referred to here as the *Baraboo interval*. Similar red strata, which are roughly equivalent, occur in other parts of the continent. Examples include the Belcher Islands in Hudson Bay, the Northwest Territories of Canada, and the Rocky Mountains of Montana and Alberta, northern Utah, and central Arizona. Collectively, such Proterozoic strata comprise the world's oldest volumetrically significant iron-oxide–bearing red beds. Conversely, red beds are not conspicuous in rocks older than about 1,800 m.y., except for the special case of local red ferric-oxide-jasperoid iron formations. Older Proterozoic (Huronian and Animikean) quartzites are, instead, either gray or white, and associated argillaceous rocks are gray or black. The color of the red quartzites is considered primary (or diagenetic) rather than due to oxidation of iron by metamorphism because of the consis-

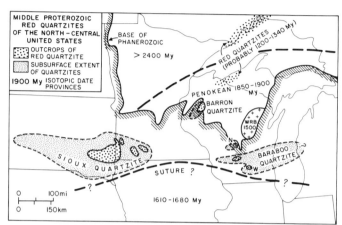

Figure 1. Present distribution of Proterozoic red quartzites in relation to other major Prepaleozoic tectonic features in the north-central United States. (N, Necedah, Wisconsin; W, Waterloo, Wisconsin; WRB, Wolf River batholith.)

tent color contrast just noted, the fact that unmetamorphosed examples are just as red as the metamorphosed ones, and because much of the interstitial iron oxide was engulfed by diagenetic quartz overgrowths (Fig. 2). Moreover, iron-oxide minerals show only minor recrystalliza-

Figure 2. Interstitial iron oxide in part engulfed within diagenetic quartz overgrowths, which demonstrates an early origin of red color. Note well-rounded grains. (Unmetamorphosed Barron Quartzite, Wisconsin; uncrossed nicols, 90 X.)

tion in metamorphosed banded iron formations. In the absence of carbon and sulphur, such minerals are exceptionally stable chemically through at least the sillimanite grade (C. Klein and C. V. Guidotti, 1974, oral commun.) Together with other well-known geochemical evidence, the early red beds have great global significance as empirical evidence of considerable free oxygen in the atmosphere—perhaps approaching half of its present abundance.

The red quartzites also are potentially important for insights into Proterozoic weathering, erosion, and depositional processes. Intuitively, we would expect important differences in such processes in the absence of land vegetation and with less atmospheric free oxygen than now. Texturally, the quartzites are consistent with a predominance of braided fluvial and eolian processes on land areas. We find some surprises, however, in the prevalence of very aluminous phyllosilicate minerals and a paucity of feldspar indicative of extremely mature chemical weathering, which is difficult to imagine without vegetation to hold the soil.

Tectonically, the quartzites are enigmatic because they speak sedimentologically of great stability and multicycling of durable constituents, as do their Paleozoic counterparts, yet some have suffered orogenic-style deformation and metamorphism. The substantial thickness of several examples implies deposition on a passive continental margin, but their deformation implies subsequent compression as if by lithosphere plate collision.

STRATIGRAHIC RELATIONSHIPS AND AGE

Red quartzitic sedimentary rocks occur in four principal areas of the north-central United States (Fig. 1). The Baraboo, Sioux, and Barron quartzites are so strikingly similar that they are correlated with considerable confidence (Dott and Dalziel, 1972). This *red trio* is the main subject of discussion, but there are also similar red quartzites on both the north and south sides of Lake Superior that appear to be younger. (Fig. 1, north edge).

Baraboo Quartzite

The Baraboo Quartzite, which has been studied most intensively, crops out in an elliptical inlier around Baraboo in southern Wisconsin (Fig. 1). It is 1,500 m thick, stratigraphically overlies rhyolitic volcanic rocks (Fig. 3), and is in turn overlain conformably by the black Seely Slate (100 m), Freedom Dolomite with iron formation (300 m), Dake Quartzite (65 m), and Rowley Creek Slate (45 m) (Fig. 4). This entire Proterozoic sequence has been folded and faulted into a complex synclinorium. Around Waterloo, Wisconsin, 50 miles southeast of the Baraboo syncline, red quartzite also crops out. This rock is lithologically identical with that at Baraboo except for a somewhat higher grade of metamorphism and more muscovite; it, too, occurs in a large synclinal structure but is less completely exposed. Drilling indicates that red quartzite occurs widely in the subsurface of southern Wisconsin (Fig. 1).

Pre-Baraboo rhyolitic volcanic rocks are mostly ash-flow tuffs and breccias, which are widespread in south-central Wisconsin (Asquith, 1964; Dott and Dalziel, 1972; E. Smith, 1978). At Baraboo, a Rb-Sr total-rock isochron of 1,640 ± 40 m.y. was obtained for them (Dott and Dalziel, 1972), but U-Pb dating of zircon farther northeast points to a real age of 1,760 ± 10 m.y. for rhyolites and associated

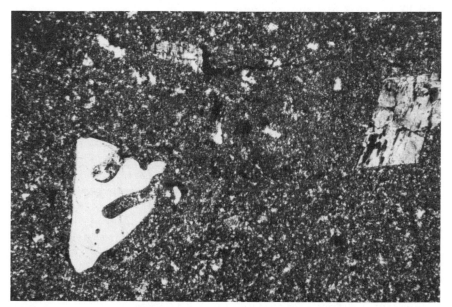

Figure 3. Porphyritic rhyolite or dacite with feldspar and embayed quartz phenocrysts in a chertlike, devitrified-glass groundmass. (Utley, Wisconsin; crossed nicols, 60 X.)

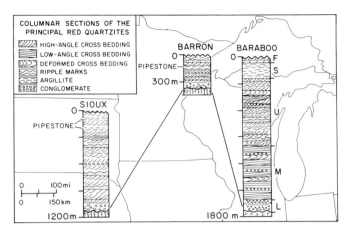

Figure 4. Generalized stratigraphic columns for the red trio (Sioux, Barron, Baraboo quartzites, principal subjects of this paper). (For Baraboo column: F, Freedom Dolomite; S, Seely Slate; L, M, U, informal subdivisions of the Baraboo Quartzite after Henry, 1975. Sioux partly after Baldwin, 1951. Barron partly after Utzig, 1972.)

granitic plutons (Van Schmus and others, 1975a; Van Schmus, 1980). Because an Rb-Sr resetting event has been recognized by Van Schmus (1980), the older U-Pb date is now regarded as more accurate.

A pegmatite dike cutting the red quartzite near Waterloo yielded an Rb-Sr date of 1,440 m.y. (Bass *in* Aldrich and others, 1959), and muscovite from a phyllite within that quartzite yielded a minimum K-Ar date of 1,410 m.y. (Goldich and others, 1966). The large Wolf River batholith complex in central and northeastern Wisconsin (Fig. 1) yields an age of 1,485 ± 15 m.y. (Van Schmus and others, 1975b). There was a widespread 1,400 to 1,500-m.y. thermal event accompanied by extensive rhyolitic volcanism and scattered plutonism across much of the midcontinent. These events, however, are thought to have been anorogenic in character (J. L. Anderson, 1983) and thus considerably younger. Van Schmus (1980) suggested that deformation and metamorphism of the red quartzites occurred instead during a 1,615 to 1,630-m.y. Rb-Sr resetting (metamorphic) event that is recognized in northern Wisconsin and adjacent parts of Michigan and Minnesota. Foliated granitic rocks 1,625 ± 25 m.y. old (U-Pb) known in northern Kansas and Missouri also are thought to reflect a postulated orogenic event in the north-central states (Van Schmus and Bickford, 1981).

We see that deposition and deformation with metamorphism of the Baraboo sequence occurred sometime between 1,450 and 1,750 m.y., which is here called the *Baraboo interval* (Table 1). On the basis of the presence of banded iron formation and overall lithology, the sequence was long ago considered to be of Huronian age (2,150 to 2,400 m.y.) and later of Animikean (1,900 to 2,000 m.y.) age. The stratigraphic relationship with the rhyolites and isotopic dating, however, indicate that it is younger than

both of them (Dott and Dalziel, 1972). Even more surprising was the realization that the Baraboo strata apparently postdate the major Penokean orogeny, which affected practically all of the Great Lakes region between 1,760 and 1,860 m.y. ago (Van Schmus, 1980). The pre-Baraboo rhyolites (1,760 m.y.) are now considered by Van Schmus to be late or post-Penokean. A long period of weathering and erosion must have followed Penokean mountain building and rhyolitic volcanism before Baraboo sedimentation commenced in order to account for leveling of mountainous topography and concentration of an enormous volume of pure quartz sand. A reasonable assumption of about 80 to 100 m.y. for such erosion would imply an age span for the entire sedimentary sequence at Baraboo of the order of 50 m.y. if Van Schmus is correct that deformation occurred 1,615 to 1,630 m.y. ago.

Sioux Quartzite

The Sioux Quartzite crops out in a series of inliers in southwestern Minnesota and adjacent South Dakota and Iowa; it is known to be widespread in the subsurface as well (Fig. 1). It is at least 1,000 m and possibly 3,000 m thick, but no stratigraphic top is known, unlike the Baraboo Quartzite (Baldwin, 1951) (Fig. 4). The Sioux was folded and metamorphosed, but less intensely than the Baraboo sequence.

The Sioux Quartzite overlies unconformably a com-

TABLE 1. ISOTOPIC DATING CONSTRAINTS ON AGES OF THE RED QUARTZITES

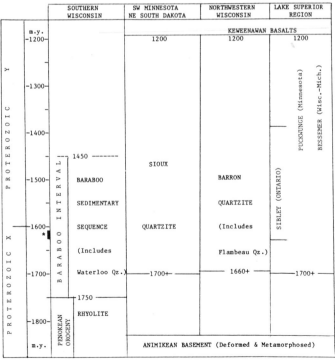

*1615-1630-m.y. metamorphic event recognized in northern Wisconsin and granites of Missouri and Kansas that probably postdate deposition of the quartzites.

plex of rocks that were last metamorphosed around 1,700 to 1,850 m.y. and were intruded by scattered 1,800 to 1,900-m.y.-old plutons (Peterman, 1966; Goldich and others, 1970). Silicic volcanic rocks in southeastern South Dakota that have minimum ages of 1,640 to 1,680 m.y. (Goldich and others, 1966) are probably older than the Sioux and may be related to the pre-Baraboo rhyolites of Wisconsin. A K-Ar date of 1,200 m.y. for micaceous argillite within the formation provides a minimum age for metamorphism of the Sioux (Goldich and others, 1961). A total-rock Rb-Sr minimum age of 1,470 to 1,530 m.y. (depending upon decay constant used) was obtained for rhyolite closely associated with Sioux Quartzite in a drill hole in Iowa (Lidiak, 1971; Dott and Dalziel, 1972). The relative age of this rhyolite is not clear, but very likely it was intrusive. In any case, the date falls within the *Baraboo interval* (1,450 to 1,750 m.y.). It follows from the dating evidence that Sioux deposition occurred sometime between 1,200 and 1,850 m.y. ago, but more probably between 1,530 and 1,750 m.y. (Table 1).

Barron Quartzite

The Barron Quartzite is less extensive in outcrop and is thinner than either the Baraboo or Sioux quartzites (Fig. 4). The main outcrop area shows less deformation and no metamorphism (Fig. 2), but an outlier of presumably correlative quartzite (known locally as Flambeau) is disturbed to about the same degree as the other two. The Barron is lithologically identical with the Baraboo and Sioux, but dating is less well constrained (Table 1). It unconformably overlies deformed and metamorphosed Animikean rocks, which farther east have yielded minimum ages of 1,660 to 1,720 m.y. (Goldich and others, 1961). These reflect a post-Penokean (but pre-Barron) metamorphic event, for plutons emplaced in northern Wisconsin during the Penokean orogeny yield dates of 1,830 to 1,850

m.y. (Van Schmus, 1980). The Barron is directly overlain by Cambrian sandstones and glacial drift. It is in close proximity to 1,100 to 1,200-m.y. Keweenawan basalts and appears to project structurally beneath them, but the Lake Owen fault may extend between the two.

Quartzites Around Lake Superior

The Puckwunge (Minnesota), Sibley (Ontario), and Bessemer (Michigan-Wisconsin) quartzites are known to underlie Keweenawan basalts and to overlie unconformably either Animikean or Archean rocks in the Lake Superior region (Fig. 1). Their stratigraphic relationships place all three within the broad age range of 1,200 to 1,700 m.y., and they were suggested to be correlative with the Baraboo-Sioux-Barron red trio by Mattis (1971), Dott and Mattis (1972), and Dott and Dalziel (1972). The Sibley in Ontario is reported to be $1,339 \pm 33$ m.y. old by a Rb-Sr whole-rock isochron for shale and dolomite (Franklin and others, 1980). The other two are less well dated (Table 1). Differences of magnetic polarity and of color and induration suggest that they are not all exactly correlative with one another, but Ojakangas and Morey (1982) regard them all as between 1,100 and 1,340 m.y. old. Contrary to earlier suggestions, they do not seem correlative with the older trio to the south.

SEDIMENTOLOGY

Composition and Provenance

Overall, quartz constitutes about 95% of the detrital sand grains, chert or jasper about 4%, and feldspar and heavy accessory minerals less than 1% each (Table 2). The heavy mineral suite is characterized by the stable species zircon, metallic oxides (chiefly magnetite and ilmenite), ru-

TABLE 2. MODAL COMPOSITIONAL DATA FOR THREE RED QUARTZITES

	QUARTZ			CHERT	FELDSPAR	HEAVY	MATRIX AND CEMENT				PHYLLOSILICATE MINERALS
	UNIT	POLYCRYS-TALLINE	FOLIA-TED	(Jasper)		MINERALS	PHYLLO-SILICATE	QUARTZ GROWTHS	CARBON-ATE	HEMATITE	KNOWN TO BE PRESENT
BARABOO (N=33)	63	3.5	0.5	4	--	1	17	10	--	0.5	pyrophyllite, kaolinite, 2M$_1$ muscovite
	Total Quartz = 67%						Total = 28%				
BARRON (N=10)	85	3-4	1	5 - 10	--	1	?	?	--	?	kaolinite, muscovite
	Total Quartz = 90%										
SIOUX	c	c	r	r	vr	vr	c	c	--	r	pyrophyllite, 2M$_1$ muscovite, kaolinite?

Sources as follows: Baraboo: Dalziel and Dott (1970); Wanenmacher (1932); X-Ray data courtesy S. W. Bailey. Barron: Hotchkiss (1915); Palmer (1940); Univ. Wisconsin thin sections; X-Ray data courtesy S. W. Bailey and Gregory Utzig. Sioux: Berg (1938); Baldwin (1951); Miller (1961); Univ. Wisconsin thin sections; X-Ray data courtesy S. W. Bailey.

c = common; r = rare; vr = very rare

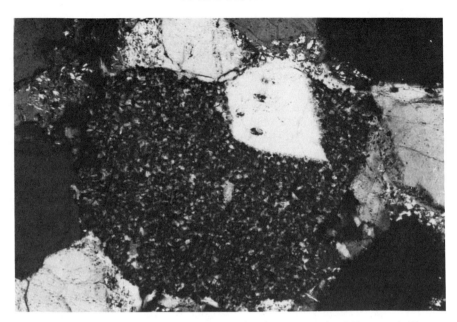

Figure 5. Porphyritic felsic volcanic fragment in Sioux Quartzite (Minnesota). Note embayed quartz phenocryst and chertlike appearance of the devitrified-glass groundmass (compare Fig. 3). (Crossed nicols, 115 X.)

tile, and apatite. Tourmaline and garnet, although present as traces in some of the formations, are conspicuously absent from the Baraboo. Phyllosilicate cement or matrix comprises typically 5% to 20%, and silica cement makes up 5% to 10% by volume. Phyllosilicate matrix is particularly conspicuous in the Baraboo Quartzite, wherein values as high as 30% were reported by Henry (1975). As a result, many Baraboo samples are quartz wackes rather than arenites. The red color is caused by a small quantity of finely disseminated interstitial hematite (Fig. 2). Chemical analyses indicate 95% to 99% SiO_2, 0.3% to 2.0% Al_2O_3 and 0.1% to 1.2% Fe_2O_3.

The detrital grains tell much about provenance. Conglomerate pebbles and granules are about 90% white quartz and 5% to 10% chert (jasper), minor slate and quartzite clasts, and a few red argillaceous intraclasts. Plutonic igneous pebbles are practically unknown, but chertlike devitrified felsic volcanic grains are present (Fig. 5). Minor hematitic chert (jasper) sand-size grains as well as pebbles attest to iron-formation sources; peloidal and oolitic textures are seen in some of these chert grains in thin section. Very rare quartz grains possessing abraded overgrowths and iron-cemented sandstone clasts indicate recycling from older sedimentary sources. Most of the nonfoliated polycrystalline sand-size grains, especially those with bubbles and minute crystal inclusions, probably were derived from quartz veins, as were the milky quartz pebbles. Slate and quartzite pebbles together with rare foliated-quartz sand-size grains indicate some metamorphic sources (Fig. 6).

Conventional wisdom attributes most of the quartz in compositionally mature sandstones and quartzites to granitic plutonic ultimate sources. Such inference then carries the very significant implication of exposure of large areas of deeply eroded continental crust as in stabilized cratons. But each of the red trio contains some embayed quartz grains derived from silicic volcanic source rocks (Fig. 7). The widespread rhyolites of central Wisconsin, which directly underlie the Baraboo Quartzite and contain many embayed quartz phenocrysts, lead me to conclude that volcanic as well as plutonic and sedimentary rocks were important sources of quartz.

The phyllosilicate mineral suite is unusual in that it is dominated by the very aluminous species pyrophyllite and kaolinite and is relatively impoverished of potassium-bearing species like muscovite (Table 2). Both the Sioux and Baraboo quartzites contain almost exclusively pyrophyllite ($Al_2Si_4O_{10}(OH)_2$). It is especially characteristic of the massive red pipestone layers in the Sioux, which according to legend, were colored by the blood of aboriginal ancestors. Pyrophyllite probably developed by metamorphism of original kaolinite ($Al_2Si_2O_5(OH)_4$). Velde (1968) argued that pyrophyllite would be favored over illite and chlorite if organic matter were absent and ferric iron were abundant, as was the case for the red trio. Experimental work by Roy and Osborn (1954) and Carr and Fyfe (1960) bears out such a reaction. The quartzite at Waterloo, Wisconsin, suffered higher grade metamorphism, as reflected by the presence of andalusite; muscovite also is more common there than at Baraboo (Geiger and others, 1982). The Barron Quartzite, which suffered the least metamorphism of the trio, contains kaolinite rather than pyrophyllite together with minor muscovite.

Figure 6. Foliated-quartz sand grain in Sioux Quartzite (Minnesota), which indicates some metamorphic sources for the red sandstones. (Crossed nicols, 36 X.)

Figure 7. Well-rounded and embayed quartz grain derived from a silicic volcanic source (compare Figs. 3 and 5). (Barron Quartzite, Wisconsin; crossed nicols, 36 X.)

The presence of very aluminous minerals in the metasedimentary rocks suggests derivation of the original clay minerals from maturely weathered soils like laterite and bauxite. The almost complete absence of detrital feldspar may also point to intense chemical weathering, for K-feldspar should have been more stable than today without land plants to cause potassium chelation (Basu, 1981). It may be that some feldspar was destroyed during metamorphism, but in that case potassium should reappear in metamorphic muscovite; only at Waterloo is such muscovite

conspicuous. While it is not difficult to imagine the general breakdown of feldspathic volcanic, older sedimentary, metamorphic, and plutonic rocks under such conditions, it is difficult to imagine extremely mature weathering of large land areas without vegetation to stabilize the soil profile. Very low topographic relief and a warm, humid climate seem indicated. On a regional basis, however, there must have been sufficient relief and gradient to allow fluvial transport of large volumes of sand. Harms (1979) has documented a similarly mature Cretaceous fluvial quartz sandstone sequence with thin, very aluminous shales in Egypt; it also formed under conditions of low relief and a humid, tropical climate.

Texture

Moderately well sorted sandstone (now quartzite) makes up about 85% to 90% of each formation (Fig. 2); conglomerate comprises 5% to 10%; red phyllite and very argillaceous fine quartzite make up the remaining 5% to 10%. Conglomerate is more abundant in the lower portions, whereas argillaceous lithologies are slightly more abundant in the upper portions, but either rock can occur at any level.

The largest pebbles in the Baraboo are about 3 cm. Maximum clasts in the Sioux are 8 cm, but in the Barron the maximum size is around 15 cm. The average size of sand in Baraboo sandstones is 0.15 to 0.30 mm or fine sand (Henry, 1975); many beds of medium and even coarse sand (0.25 to 1.0 mm) are also present. The Barron and Sioux have sand grains averaging about 0.2 to 0.5 mm or medium sand, but the total range is from about 0.1 to 2.0 mm (Baldwin, 1951; Utzig, 1972). In all cases, the larger sand grains are very well rounded, whereas finer ones are angular (Figs. 2, 7). Rounding is so good for many coarse grains as to suggest an important earlier eolian abrasive history (Kuenen, 1960). In some cases, metamorphism has produced interlocking quartz grain boundaries, but on the whole, original shapes are surprisingly little changed. Probably this is due largely to the prevalent fine matrix, which cushioned the sand grains; minutely digitated boundaries indicate some reaction between quartz and matrix. Ubiquitous undulatory extinction and moderately common deformation (Boehm) lamellae in quartz, especially in coarser, matrix-poor samples, record the strong deformation suffered by the Baraboo and Sioux quartzites (see Dalziel and Stirewalt, 1975).

Sedimentary Structures

The red trio is characterized by prominent cross bedding and to a lesser extent by asymmetrical ripple marks. Current lineation is also present. Most cross sets are 15 to 20 cm thick; the range is 1 to 100 cm, but sets more than 30 cm thick are rare. Restored dip angles range from 5° to 50°; those over 35° were oversteepened either by penecontemporaneous disturbance or by later structural deformation. *Low-angle cross bedding* (dips <15°) is common in the lower part of the Baraboo Quartzite, but otherwise *high-angle cross bedding* (dips >15°) is most characteristic of all formations. Low-angle cross laminae are as much as 3 m long in sets 10 to 30 cm thick. Inverse grading was observed in some of these low-angle laminae by Henry (1975). One-pebble-thick lags occur along some of the basal bounding

surfaces. Most individual cross laminae appear to be more or less planar; however, trough-shaped ones also are present. Trough sets probably are more abundant than they seem because the two-dimensional nature of many outcrops inhibits their recognition. Some cross laminae show textural grading from coarse sand or granules in the lower part of a lamina to relatively fine sand upward. Such graded cross sets develop when poorly sorted grains avalanche down lee faces of bedforms. The geometry of sets of cross laminae is predominantly tabular because prominent bounding surfaces between sets are essentially planar and parallel. Some of these bounding surfaces are mantled by argillaceous (now phyllitic) material only 1 to 2 cm thick. A few *reactivation surfaces* have been observed near the top of the Baraboo Quartzite. Penecontemporaneously deformed cross bedding is known in the Baraboo Quartzite at several localities. *Recumbent* or *overturned cross laminae* occur at many positions, and disharmonically *contorted cross laminae* occur in the upper half of the formation.

Asymmetrical ripple marks are prominent on many bedding surfaces in all of the quartzites, but especially in the upper half of the Baraboo (Fig. 4). The ripple crests are somewhat sinuous in plan, but are not extremely so. The average ripple amplitude is about 1 cm, and average wavelength is between 6 and 12 cm, resulting in a typical ripple index <10 (R.I. = wave-length/amplitude). Henry (1975) noted that in the Baraboo Quartzite ripples tend to occur repetitively at the top of an interval of cross bedding; a thin phyllite seam mantles the ripples. Orientations of the ripples vary considerably (Baldwin, 1951; Brett, 1955; Dott and Dalziel, 1972).

The *red argillaceous layers* generally range from <1 cm to 1 m in thickness, but one 10-m layer is reported in the Sioux (Baldwin, 1951). In the Baraboo Quartzite they are now phyllites with prominent cleavage due to structural deformation, but in the Sioux and Barron, where they are less deformed, mudstones are either massive or show faint parallel lamination. Rarely ripples on fine sand laminae may be interstratified with the mudstones. Polygonal "mudcracks" are reported in the Sioux (Baldwin, 1951).

Sediment Dispersal Patterns

Orientation analyses have been made extensively of current-formed sedimentary structures for all of the quartzites. Figure 8 summarizes the results in simplified form. Dispersion or scatter is considerable in all of the raw data, be it for individual cross-set orientations, trough-set axes, or ripple crest trends. For example, locality standard deviations for cross-set orientations vary from ±28° to ±118°, but slightly over half of the localities show less than ±60°. The mean directions show considerable variation from area to area (Dott and Dalziel, 1972) but, nonetheless, display an overall net transport of detritus toward the southerly

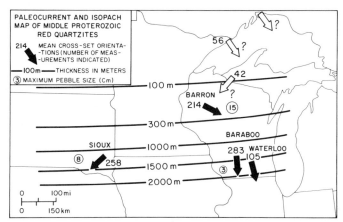

Figure 8. Generalized paleocurrent and present-thickness map for the Proterozoic red quartzites of the north-central United States. Maximum pebble sizes also shown (circles). Solid arrows for red trio; open arrows for younger Lake Superior quartzites. (Cross bedding orientations generalized from Brett, 1955; Mattis, 1971; Dott and Dalziel, 1972.)

half of the compass (in present coordinates). The gradient of maximum clast sizes noted above from 15-cm cobbles in northwestern Wisconsin (Barron) to 3-cm pebbles 300 km to the southeast at Baraboo and mean sand sizes from medium sand in the north to fine sand at Baraboo also support a general north-to-south dispersal. It is noteworthy that the gross dispersal pattern was approximately in the direction of apparent thickening of the sedimentary wedge (Fig. 8). Also noteworthy is a complete lack of known bimodal patterns, such as are commonly associated with tidal deposits.

Processes and Environments of Deposition

Until recently, a shallow nearshore marine environment for most if not all of the red quartzites was tacitly assumed (e.g., Dalziel and Dott, 1970; Dott and Dalziel, 1972). As Henry (1975) noted, however, in the complete absence of fossil evidence, that assumption is tenuous, and fluvial deposition seems equally plausible. Moreover, it taxes credulity to imagine that any single environment persisted at one locale during the accumulation of sequences as thick as the Baraboo and Sioux. Also, it is impossible to argue that the same shoreline environment existed simultaneously for deposition of, say, the Baraboo and Barron quartzites, which are separated by 300 km in the direction of regional sediment dispersal. Only if the formations are quite diachronous, which is possible, could they both be shoreline deposits.

What can be inferred in the absence of fossil evidence? Marine deposition seems more definite for the Seely Slate and Freedom Dolomite overlying the Baraboo Quartzite (Fig. 4). The Freedom is not only carbonate-rich but also contains a banded iron formation and polygonal mud-cracks indicative of alternate submergence and emergence.

The thick argillaceous Seely is black and contains chlorite and muscovite, unlike the red argillaceous strata. If these upper units are accepted as marine, could the entire Baraboo sequence be transgressive? That the Baraboo Quartzite was an aqueous deposit cannot be questioned because much of it is coarser and less well sorted than most eolian deposits and the ripple index is too low ($\leqslant 10$) for eolian ripples ($\geqslant 20$). Finally, the recumbent (overturned) cross laminae required flowing water for their formation (Doe and Dott, 1980).

The sedimentologic characteristics of all of the quartzites are most consistent with braided fluvial and/or littoral and inner shelf deposition. Pebbles in the Baraboo would require near-bottom threshold water velocities of the order of 150 to 200 cm/s; coarser ones in the Sioux and Barron would require velocities of 300 to 400 cm/s. Typical sands of the red quartzites could be moved by flows with near-bottom velocities of the order of 20 to 40 cm/s. Flow velocities of these magnitudes are possible in both rivers and shallow seas, but the maximum values would be achieved in the sea only by rare storm conditions. Considerable variations of flow are indicated by interstratified pebbles, sand, and argillaceous sediments, but these could occur in either environment.

The prevalence of planar cross bedding suggests that the most frequent bedform was relatively straight-crested dunes or sand bars with the trough sets indicating more irregular dunes. Slightly sinuous ripple marks were superimposed upon these as second-order forms. Such composite bedforms are produced by moderately intense flow conditions (lower regime) with velocities of 20 to 100 cm/s. The planar bounding surfaces that truncate cross sets and the one-pebble-thick conglomerate layers reflect more intense (upper regime) sheet flow with considerable scouring of the bedforms, as well as relatively rare transport and winnowing of coarse material. Velocities would have been 100 to 500 cm/s. As flow waned, dunes and ripples reformed and buried the scoured and winnowed surfaces. Occasionally when flow intensity was very low (velocity <10 cm/s), fine clays settled from suspension to mantle the sediment surface.

Lenticular conglomerate zones up to several meters thick characterize the basal parts of most of the formations. These are more likely to be fluvial than marine, for waves tend to segregate gravel into thin, well-sorted layers that are less lenticular than fluvial channel gravels (Clifton, 1973). Henry (1975) also detected fining-upward sequences within the lower pebbly Baraboo Quartzite, which he interpreted, together with the unimodal paleocurrent pattern, to be more indicative of braided fluvial deposition.

Low-angle cross sets common above the basal pebbly zone of the Baraboo Quartzite resemble beach face or swash zone lamination. Henry (1975) even found some inverse size grading in these strata like that described by Clifton (1969) as characteristic of beach lamination. It is conceivable that this interval represents an initial transgression and that the upper half of the Baraboo contains shallow marine deposits produced by vast fields of migrating ripples and dunes in continuously agitated water. But should we not expect some wave-formed symmetrical ripples in this case as well as at least some evidence of tidal currents and possibly evidence of intermittent emergence? Reactivation surfaces near the top of the formation at but one locality are the only features even suggestive of a tidal influence, but such surfaces also can form in river deposits by seasonal fluctuations of water level. Mudcracks in the Sioux indicate intermittent desiccation, but such features require longer to form than a typical six-hour low-tide cycle; they are most characteristic both of infrequently wetted supratidal environments and of rivers whose levels fluctuate. If the upper Baraboo were, indeed, marine, then its unimodal paleocurrent pattern would reflect either an asymmetrical, ebb-dominated tidal regime or a wind-driven current regime on a tideless shelf. Perhaps instead the pattern of disperal in the direction of stratigraphic thickening is in reality testimony of a fluvial origin.

Henry (1975) suggested that the entire Baraboo Quartzite could have resulted from braided fluvial deposition. Lack of obvious channel structures is not surprising because braided channels in sandy, cohesionless material are broad and shallow. Thin, lenticular sediment bodies are the typical product; the scale and lateral continuity of Baraboo strata suggest broad channels hundreds to thousands of meters in width and a few meters deep on the average. Conglomeratic layers could be easily explained as high-discharge deposits, and the argillaceous layers either as low-stage or flood overbank deposits (or both). Desiccation mudcracks would be expected in these. The dominant high-angle cross-bedded and rippled sequences could reflect migrating transverse bars with rippled tops, which are characteristic of braided rivers. Such bars typically produce tabular sets of planar cross bedding (N. Smith, 1970), whereas irregular dunes produce trough sets. The low-angle cross bedding is less easily explained by the fluvial model; presumably it would reflect low-angle lateral rather than leeward avalanche faces of bars. Deformed cross bedding, while not confined thereto, is certainly most common in fluvial deposits (Doe and Dott, 1980). The deformed Baraboo sets could represent disturbance of loosely packed, saturated sands by sudden increase of current shear over the bed or by a sudden rise of water level within loosely packed sands that had been partially exposed during low-flow stages. Rare reactivation surfaces also could reflect partial erosion of bar surfaces during low-stage exposure followed by reactivation of dunes with subsequent submergence. It is significant that the reactivation surfaces are associated closely with deformed cross bedding. Henry (1975) argued that the rather abundant phyllosilicate matrix found in most of the Baraboo sandstones is more easily explained by a fluvial than a marine environment. Ceaseless wave agitation should winnow such material efficiently, but a river with fluctuating discharge would winnow less effectively. Moreover, fine clay particles can infiltrate into a coarser sandy or gravelly river bed during falling water.

Whether or not the quartzites themselves are entirely fluvial or at least partly marine, it seems clear that the post-Baraboo slates and dolomite with iron formation are marine. The upper few hundred meters of the Baraboo, which show the maximum lateral continuity of lithologies (at least 6 km), may represent the transition to marine deposition. Broadly speaking, therefore, the southward-thickening sedimentary wedge, of which the red quartzites constitute the major share, seems to be in some sense transgressive.

TECTONIC SIGNIFICANCE

Unconformable basal contacts, great compositional maturity, and widespread distribution underscore the post-orogenic nature of the red quartzites. By analogy with better-known counterparts, I postulate a tectonically stable continental margin setting for their deposition. The quartzites appear to be remnants of a vast sedimentary wedge that blanketed at least the southern margin of the post-Penokean craton of Proto–North America. The great thickness of the Sioux and Baraboo seem to preclude an intracratonic setting. An aulacogen has been suggested, but there is no evidence for a narrow, two-sided basin, and sedimentation in such rifts generally progresses from deeper to shallower marine and even nonmarine upward—the reverse of the red quartzites. A passive continental margin is most plausible. Some combination of braided rivers in a humid climate and shallow marine processes dispersed clastic sediments across that continental margin.

The setting postulated for the Proterozoic red trio is like that already deduced for several better-known passive, Atlantic-type continental margin sequences. Examples include the late Proterozoic Beltian of Montana and Alberta and the Eocambrian-Cambrian clastics of both the Appalachian and Cordilleran regions. In all of those cases, thick wedges of clastic sediments accumulated under stable tectonic conditions, but were sooner or later caught up in compressional tectonics due to changes of lithosphere plate interactions.

So, too, the mid-Proterozoic red quartzite wedge of the north-central United States was subjected to strong compressive tectonics, which folded and metamorphosed the Baraboo and Sioux sequences. What had been a passive margin became an active one that suffered orogenesis sometime between 1,450 and 1,630 m.y. ago. The Baraboo sequence and, to a lesser extent, the Sioux Quartzite suffered intense ductile deformation, which resulted in penetrative cleavages, boudinage, tension gashes, and quartz deformation lamellae (Dalziel and Dott, 1970). Greenschist metamorphism was pervasive and amphibolite grade was reached at Waterloo, Wisconsin. The common Baraboo-Sioux assemblage pyrophyllite-quartz-(muscovite)-hematite) indicates an upper temperature limit of 420°C at 2-kbar pressure (Kerrick, 1968). The kaolinite-quartz assemblage of the Barron Quartzite indicates an upper temperature limit of 380°C at 2-kbar pressure (Carr and Fyfe, 1960). The highest grade known was developed in the Baraboo Quartzite at Waterloo, where an andalusite-staurolite-muscovite-chlorite-hematite assemblage developed (but was retrograded). This assemblage suggests an upper stability limit of 350° to 550° at <3.8-kbar pressure (Geiger and others, 1982). Post metamorphic (but pre-Paleozoic) hydrothermal activity is recorded in the Baraboo Quartzite by small quartz crystals associated with dickite and kaolinite in breccia zones and along joint surfaces. Fluid inclusions in the quartz crystals have indicated a temperature of deposition of 105° to 107°C, which is consistent with that required for the formation of dickite (S. W. Bailey, 1970, personal commun.).

The most plausible hypothesis to explain the seemingly anomalous deformation and metamorphism of the mature red quartzites deposited upon a stable, passive continental margin is either an arc-continent or continent-continent collision about 1,600 m.y. ago (Fig. 9), as was suggested a decade ago (Dott and Dalziel, 1972). Evidence for testing this hypothesis is scant because of the inaccessibility of Prepaleozoic rocks in the critical region beneath Iowa and Illinois where a suture is postulated (Fig. 1), but the case is stronger now than in 1972. What, then, can be said to support this bold speculation? *First,* most of the basement rocks from deep wells in that region as well as outcrops in Missouri and westward to Arizona yield only isotopic dates of 1,400 to 1,700 m.y., in sharp contrast with the region to the north with many older dates (Van Schmus and Bickford, 1981). *Second,* no proven granitic crust older than 1,625 m.y., such as is present in the Great Lakes region, is known to the south. *Third,* Prepaleozoic mafic rocks from a few wells in Iowa and Illinois might represent a suture zone (see Dott and Dalziel, 1972, Fig. 1). Mafic dikes or sills (now amphibolite) also occur in the Baraboo Quartzite at Waterloo (Geiger and others, 1982), in the Sioux (Baldwin, 1951), and within the central Wisconsin rhyolites (E. Smith, 1978). *Finally,* a widespread, 1,615 to 1,630-m.y. thermal (metamorphic) event is known in Wisconsin, and granitic plutons and rhyolitic volcanic rocks ranging in age from 1,610 to 1,680 m.y. occur from Missouri to Arizona (Van Schmus and Bickford, 1981).

Van Schmus (1976) has postulated northward subduction with development of an Andean-like magmatic arc on the southern edge of Proto–North America to account for the Penokean orogeny (1,760 to 1,860 m.y.) (Fig. 9). The

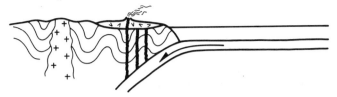

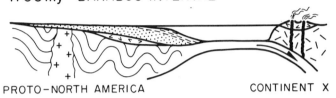

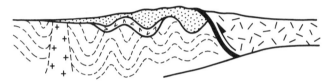

Figure 9. Hypothetical development of the southern margin of Proto–North America, 1,800 to 1,600 m.y. ago, showing deposition of red quartzites followed by deformation by postulated plate collision.

last stages of such subduction might have produced the pre-Baraboo rhyolitic volcanism in southern Wisconsin (1,760 ± 10 m.y.), but Van Schmus (1980, oral commun.) also suggested possible anorogenic melting of Penokean crust. I suggest that northward subduction then ceased, and after much erosion, the red quartzites accumulated on the now-stable and deeply eroded margin of Proto–North America. Subsequently, southward-directed subduction began to consume the sea floor lying south of that margin as either a volcanic island arc or another continent approached (Fig. 9). About 1,600 m.y. ago, that arc or continent collided with and deformed the southern margin of Proto–North America, much as occurred during Cenozoic time in the old Tethyan seaway of Eurasia. Northward subduction is ruled out for the later event because no 1,600-m.y.-old volcanic complex is known on the Proto–North American margin. Instead, the 1,610 to 1,680-m.y.-old volcanic and granitic rocks extending from Arizona to Missouri are thought to have been the products of southward subduction beneath a colliding continent; they are of the appropriate age and lie in the appropriate direction.

How could the proposed hypothesis be tested? More drill holes to Prepaleozoic basement in Iowa and Illinois and paleomagnetic data from both sides of the postulated suture offer the most obvious tests. Additional isotopic studies of basement rocks there also should help. What of an immature clastic molasse wedge that would be expected from the erosion of mountains raised by collision? If such deposits existed, they have not been recognized, but it is entirely possible that they were removed by erosion during the 300 to 400-m.y. interval between the postulated collision and Keweenawan events (Table 1). A widespread unconformity everywhere separates Keweenawan rocks from older ones, which yield isotopic dates >1,625 m.y. Local survival of 1,200 to 1,600-m.y.-old strata perhaps could be sought in poorly dated Proterozoic sequences, however, especially in the subsurface.

CONCLUSION

Quartz-bearing felsic volcanic rocks, together with older sedimentary and plutonic ones, provided much quartz sand, whereas quartz veins and chert from iron formations provided gravel, all of which accumulated to a maximum thickness of nearly 2,000 m on the southern, stable margin of Proto–North America between 1,450 and 1,750 m.y. ago—*the Baraboo interval.* Erosion and weathering of Penokean mountains for about 100 m.y. preceded deposition of the red sediments, which apparently spanned a period of the order of 50 m.y. Very mature weathering destroyed all but traces of both plagioclase and K-feldspar in the source and apparently produced virtual bauxitic soils despite the absence of land vegetation. Topographic relief apparently was modest and the climate was probably humid and warm. The soils gave rise to aluminum-rich clay sediments dominated by kaolinite, later mostly metamorphosed to pyrophyllite. The red quartzites of the north-central United States are among the oldest red beds known, and the presence of primary ferric oxide permeating all lithologies from conglomerate to claystone attests to significant free oxygen in the atmosphere by nearly 1,800 m.y. ago.

A vegetation-free landscape containing much sand would be dominated by eolian and braided fluvial processes such as are documented in the Proterozoic of northwestern Canada (Ross and Donaldson, 1982). Proterozoic eolian deposits have not been recognized in the Great Lakes region, but well-rounded quartz grains suggest a history of eolian abrasion. All of the red quartzites can be explained as braided fluvial deposits formed on a sandy coastal plain several hundred kilometers wide. In the absence of fossil evidence, however, any environmental interpretation remains somewhat permissive. The only plausible alternative to a fluvial setting is the littoral-shallow shelf environment, which could produce many features superficially identical with braided fluvial deposits. For example, both environments tend to produce tabular strata that contain very thin, broadly lenticular conglomerate layers, ripples, and high-angle cross stratification of the sort present in the quartzites. Low-angle cross bedding might be explained as beach lamination, but the abundance of phyllosilicate matrix in the sandstones seems more explicable by fluvial processes. Predominantly unimodal orientations of paleocurrent indicators also favor a fluvial origin. While the quartzites themselves remain somewhat ambiguous as to environment, the slates and dolomite with banded iron formation that overlie the Baraboo Quartzite are more confidently interpreted as marine. Therefore, at least the upper part of the sedimentary sequence seems to record a marine transgression over the southern margin of Proto–North America.

The great compositional maturity and geometry of the quartzites point to deposition on a tectonically stable or passive continental margin, yet their present structural and metamorphic character indicate profound orogenic compression. Therefore, it is postulated that the quartzites were deformed by a plate collision after southward subduction of an ocean floor that lay between Proto–North America and an approaching arc or continent. Rhyolitic and associated granitic rocks known from Missouri to Arizona are suggested as products of such subduction, and a 1,615 to 1,630-m.y. metamorphic event known in northern Wisconsin probably also resulted from the postulated collision. Thus, a hypothesis of southeastern North American continent-building that accommodates the important constraints imposed by the Proterozoic red quartaites is offered for the testing.

ACKNOWLEDGMENTS

I acknowledge the continual stimulation and encouragement of L. G. Medaris, W. R. Van Schmus, and R. W. Ojakangas in my studies of Precambrian sediments and tectonics. S. W. Bailey and C. V. Guidotti provided important advice on mineralogy. Medaris, Bailey, Guidotti, Van Schmus, and Ojakangas also criticized the manuscript. I gratefully acknowledge the thesis studies of D. M. Henry, A. F. Mattis, and C. F. Utzig, which provided important data not otherwise available in the literature.

REFERENCES CITED

Aldrich, L. T., Wetherill, G. W., Bass, M. N., Compston, W., Davis, G. L., and Tilton, G. R., 1959, Mineral age measurements, *in* Annual Report of Director of Department of Terrestrial Magnetism: Carnegie Institution of Washington Yearbook 58, p. 246–247.

Anderson, J. L., 1983, Proterozoic anorogenic granite plutonism of North America, *in* Medaris, L. G., Byers, C. W., Mickelson, D. M., and Shanks, W. C., eds., Proterozoic geology: Selected papers from an international Proterozoic symposium: Geological Society of America Memoir 161 (in press).

Asquith, G. B., 1964, Origin of the Precambrian Wisconsin rhyolites: Journal of Geology, v. 72, p. 835–847.

Baldwin, W. B., 1951, The geology of the Sioux Formation [Ph.D. dissertation]: Columbia University, 161 p.

Basu, A., 1981, Weathering before the advent of land plants: Evidence from unaltered detrital K-feldspars in Cambrian-Ordovician arenites: Geology, v. 9, p. 132–133.

Berg, E. L., 1938, Notes on Catlinite and the Sioux Quartzite: American Mineralogist, v. 23, p. 258–268.

Brett, G. W., 1955, Cross bedding in the Baraboo Quartzite, Wisconsin: Journal of Geology, v. 63, p. 143–148.

Carr, R. M., and Fyfe, W. S., 1960, Synthesis fields of some aluminum silicates: Geochimica and Cosmochimica Acta, v. 2, p. 99–109.

Clifton, H. E., 1969, Beach lamination: Nature and origin: Marine Geology, v. 7, p. 553–559.

——— 1973, Pebble segregation and bed lenticularity in wave-worked versus alluvial gravel: Sedimentology, v. 20, p. 173–187.

Dalziel, I.W.D., and Dott, R. H., Jr., 1970, Geology of the Baraboo district, Wisconsin: Wisconsin Geological and Natural History Survey Information Circular 14, 164 p.

Dalziel, I.W.D., and Stirewalt, G. L., 1975, Stress history of folding and cleavage development, Baraboo syncline, Wisconsin: Geological Society of America Bulletin, v. 86, p. 1671–1690.

Doe, T. W., and Dott, R. H., Jr., 1980, Genetic significance of deformed cross bedding—with examples from the Navajo and Weber Sandstones of Utah: Journal of Sedimentary Petrology, v. 50, p. 793–812.

Dott, R. H., Jr., and Dalziel, I.W.D., 1972, Age and correlation of the Precambrian Baraboo Quartzite of Wisconsin: Journal of Geology, v. 80, p. 552–568.

Dott, R. H., Jr., and Mattis, A. F., 1972, A post-Animikean–pre-Keweenawan transgressive sand blanket over the Lake Superior region [abs.]: Program for Annual Meeting of Geological Society of America, p. 490–491.

Franklin, H. M., McIlwaine, W. H., Poulsen, K. H., and Wanless, R. K., 1980, Stratigraphy and depositional setting of the Sibley Group, Thunder Bay, Ontario, Canada: Canadian Journal of Earth Sciences, v. 17, p. 633–651.

Geiger, C., Guidotti, C. V., and Petro, W., 1982, Some aspects of the petrologic and tectonic history of the Precambrian rocks at Waterloo, Wisconsin: Geoscience Wisconsin: v. 6, p. 21–38.

Goldich, S. S., Nier, A. O., Baadsgaard, H., Hoffman, J. H., and Krueger, H. W., 1961, The Precambrian geology and geochronology of Minnesota: Minnesota Geological Survey Bulletin 41, 193 p.

Goldich, S. S., Lidiak, E. G., Hedge, C. E., and Walthall, F. G., 1966, Geochronology of the midcontinent region, United States: Journal of Geophysical Research, v. 71, p. 5375–5408.

Goldich, S. S., Hedge, C. E., and Stern, T. W., 1970, Age of the Morton and Montevideo gneisses and related rocks, southwestern Minnesota: Geological Society of America Bulletin, v. 81, p. 3671–3695.

Harms, J. C., 1979, Alluvial-plain sediments of Nubia, Southwestern Egypt [abs.]: American Association of Petroleum Geologists Bulletin, v. 63, p. 829.

Henry, D. M., 1975, Sedimentology and stratigraphy of the Baraboo Quartzite of south-central Wisconsin [M.S. thesis]: University of Wisconsin, 90 p.

Hotchkiss, W. O., 1915, Mineral land classification in parts of northwestern Wisconsin: Wisconsin Geological and Natural History Survey, Bulletin XLIV, 378 p.

Kerrick, D. M., 1968, Experiments of the upper stability limit of pyrophyllite at 1.8 kilobars and 3.9 kilobars water pressure: American Journal of Science, v. 266, p. 204–214.

Kuenen, Ph. H., 1960, Experimental abrasion: 4 Eolian action: Journal of Geology, v. 68, p. 427–449.

Lidiak, E. G., 1971, Buried Precambrian rocks of South Dakota: Geological Society of America Bulletin, v. 82, p. 1411–1420.

Mattis, A. F., 1971, Lower Keweenawan sediments of the Lake Superior region: Proceedings of 17th Annual Institute on Lake Superior Geology, Duluth, p. 45–46 (also M.S. thesis, University of Minnesota-Duluth).

Miller, T. P., 1961, A study of the Sioux Formation of the New Ulm area [M.S. thesis]: University of Minnesota, 75 p.

Ojakangas, R. W., and Morey, G. B., 1982, Keewenawan pre-volcanic quartz sandstones and related rocks of the Lake Superior region: Geological Society of America Memoir 156, p. 85–96.

Palmer, H. A., 1940, Correlation of the Barron Quartzite [M.A. thesis]: University of Wisconsin, 40 p.

Peterman, Z. E., 1966, Rb-Sr dating of middle Precambrian metasedimentary rocks in Minnesota: Geological Society of America Bulletin, v. 77, p. 1031–1044.

Ross, G. M., and Donaldson, J. A., 1982, A Proterozoic aeolian sand sea in the Hornby Bay Group, Northwest Territories, Canada [abs.]: Abstracts for 11th International Sedimentological Congress, Hamilton, Ontario, p. 68.

Roy, R., and Osborn, E. F., 1954, The system Al_2O_3 - SiO_2 - H_2O: American Mineralogist, v. 39, p. 853–885.

Smith, E. I., 1978, Precambrian rhyolites and granites in south-central Wisconsin: Field relations and geochemistry: Geological Society of America Bulletin, v. 89, p. 875–890.

Smith, N. D., 1970, The braided stream depositional environment: Comparison of the Platte River with some Silurian clastic rocks, north-central Appalachians: Geological Society of America Bulletin, v. 81, p. 2993–3014.

Utzig, G. F., 1972, Comparative petrographic studies of the Barron and Flambeau Quartzites [Senior Honors Thesis]: University of Wisconsin, 19 p.

Van Schmus, W. R., 1976, Early and Middle Proterozoic history of the Great Lakes areas, North America: Philosophical Transactions of the Royal Society of London, v. 280A, p. 605–628.

——— 1980, Chronology of igneous rocks associated with the Penokean orogeny in Wisconsin, *in* Morey, G. B., and Hanson, G. N., eds., Selected studies of Archean gneisses and Lower Proterozoic rocks, southern Canadian Shield: Geological Society of America Special Paper 18, p. 159–168.

Van Schmus, W. R., and Bickford, M. E., 1981, Proterozoic chronology and evolution of the midcontinent region, North America, *in* Kröner, A., ed., Precambrian plate tectonics: New York, Elsevier, p. 261–296.

Van Schmus, W. R., Thurman, M. E., and Peterman, Z. E., 1975a, Geology and Rb-Sr chronology of middle Precambrian rocks in eastern and southern Wisconsin: Geological Society of America Bulletin, v. 86, p. 1255–1265.

Van Schmus, W. R., Medaris, L. G., Jr., and Banks, P. O., 1975b, Geology and age of the Wolf River batholith, Wisconsin: Geological Society of America Bulletin, v. 86, p. 907–914.

Velde, B., 1968, The effect of chemical reduction on the stability of pyrophyllite and kaolinite in pelitic rocks: Journal of Sedimentary Petrology, v. 38, p. 13–16.

Wanenmacher, J. M., 1932, Paleozoic strata of the Baraboo area, Wisconsin [Ph.D. thesis]: University of Wisconsin, 104 p.

MANUSCRIPT ACCEPTED BY THE SOCIETY MARCH 4, 1983

Typeset by WESType Publishing Services, Inc., Boulder, Colorado
Printed in U.S.A. by Malloy Lithographing, Inc., Ann Arbor, Michigan

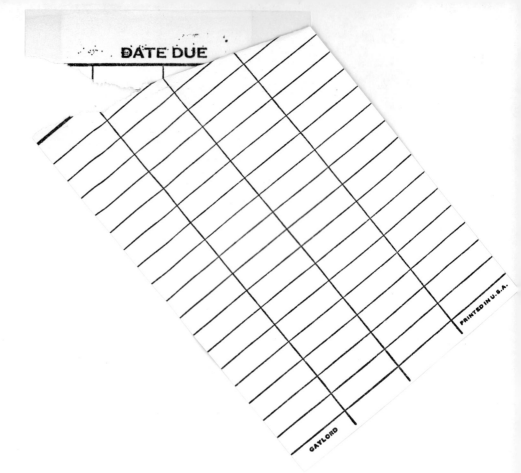

DATE DUE

GAYLORD   PRINTED IN U.S.A.